Studies in Computational Intelligence 419

Editor-in-Chief

Prof. Janusz Kacprzyk
Systems Research Institute
Polish Academy of Sciences
ul. Newelska 6
01-447 Warsaw
Poland
E-mail: kacprzyk@ibspan.waw.pl

For further volumes:
http://www.springer.com/series/7092

Joanna Kołodziej

Evolutionary Hierarchical Multi-Criteria Metaheuristics for Scheduling in Large-Scale Grid Systems

Author
Joanna Kołodziej
Cracow University of Technology
Cracow
Poland

ISSN 1860-949X e-ISSN 1860-9503
ISBN 978-3-642-28970-5 e-ISBN 978-3-642-28971-2
DOI 10.1007/978-3-642-28971-2
Springer Heidelberg New York Dordrecht London

Library of Congress Control Number: 2012934503

Printed on acid-free paper

Springer is part of Springer Science+Business Media (www.springer.com)

To my Family and Friends for Their Love and Support

Foreword

Emerging paradigms for the development and deployment of massively distributed computational systems allow resources to span diverse locations, organizations, and platforms connected through wide area networks. In such systems, both service provision and services may arrive, be organized, and dissipated, as computational capabilities are formulated and reformulated without reference to any central authority or any coordinator.

The term *Grid* was coined in the mid-1990s in reference to technologies that would allow consumers to utilize computing power on demand. Ian Foster, in his pioneering work on grid computing, posited that a simple grid environment may be created by standardizing the protocols used to request computing power across several computing clusters. His methodology was analogous in form and utility to the conventional electric power grid. Although the concept of Grid Computing has grown far beyond its original intent, from an engineering perspective grid scheduling retains the original (and general) objective of system resource allocation. A well-known example of the original objective is *load balancing*, in which the task of providing a resource is distributed evenly between some nodes. However, more complex objectives are defined in today's grid systems, such as. both stable and uneven allocations where other factors are taken into account, including users' preferences related to quality of service issues, underlying computational overhead incurred by the service provider, security of access to the resources, energy utilized by the system, and many others. The ability to effectively allocate resources in a desired configuration in a scalable and robust manner is essential.

This book presents a new categorization of grid scheduling problems. Two new scheduling criteria, namely security and energy consumption, are embedded in the proposed scheduling models. In addition, the author demonstrates that the grid scheduling problem may be interpreted as a difficult decision problem for grid users working at the different grid levels. The fundamental features arising in user behavior in grid scheduling include: dynamics, selfishness, cooperativeness, trustfulness, and their symmetric and asymmetric roles. User decisions are modeled on game-theoretical models. All of the issues identified above provide a basis upon which traditional scheduling problems may be viewed from a contemporaneous and unique perspective.

Two major challenges in the use of evolutionary-based techniques for solving dynamic optimization problems are (1) to generate and maintain sufficiently high diversity levels in the population, and (2) to evolve robust solutions that track the optimal solutions identified during the process. Ideally, we want an adaptive algorithm that responds in an appropriate way every time a change in the environment occurs. This book presents a generic model for a hierarchical multi-population genetic scheduler that enables an undemanding configuration of the numerous genetic operators and an effective exploration of the search space with an adaptive accuracy. This model may be easily adapted to a range of scheduling scenarios. The functionality of this model and its effectiveness in multi-criteria grid scheduling is justified in the comprehensive experimental analysis.

I believe that you will find the contributions in this book very interesting as they provide innovative and contemporary expositions of new concepts and techniques in advanced scalable grid computing. It is devoted to the study of common and related subjects in two intensive research areas of distributed computing and evolutionary computation. It is a very timely volume to be welcomed by the wider Computational Intelligence community and beyond.

Adelaide, Australia

Zbigniew Michalewicz
January 2012

Preface

In the recent years, we are witnessing a growing interest in the need for designing intelligent models and methodologies to address and solve complex issues within the domain of large-scale distributed systems that provide high performance capabilities to a wide range of applications with different, and at times conflicting requirements. In today's Computational Grids, Clouds, or modern Clusters the Information Technology (IT) resources usually belong to different owners (institutions, enterprises or individuals) and are managed by different administrators. Resource administrators conform to different sets of rules and configuration directives, and can impose different usage policies on the system users.

Highly complex large-scale distributed computing systems, which could be made up of hundreds or thousands of various components (computers, databases, etc) must provide (in fact) a wide range of services and high performance computing platforms. Moreover, a user in one locality (geographical or managerial) might not be able to have control over other parts of the system. Various types of information and data processed in the large-scale dynamic environment may be incomplete, imprecise, fragmentary and overloading, which complicates the specification of proper evaluation criteria, assignment scores, availability of resources, and the final collective decisions of the users. The system complexity may also be the reason of the higher energy consumption. All of the above mentioned issues will necessitate the development of intelligent resource management techniques.

Scheduling problems in highly heterogeneous environments may be considered as a family of NP-complete optimization problems. Depending on the restrictions imposed by the application needs, the complexity of the problem can be determined by the number of objectives to be optimized, such as (single vs. multi-objective), the type of the environment (static vs. dynamic), the processing mode (immediate vs. batch), and tasks interrelations (independence vs. dependency). Therefore, there is a great need for the development of newer highly scalable scheduling models with newer metrics for large-scale distributed systems that could capture all of the complexity and provide meaningful measures for a wide range of applications. That is to say that new classes of algorithms, schedulers and simulation models, that would be able to characterize the system dynamics and variety of services and applications, must be developed.

Artificial Intelligence-based metaheuristics, such as fuzzy logic, neural networks, evolutionary and memetic algorithms have shown great potential to solve many demanding, real-world decision and optimization problems in uncertain large-scale environments. Heuristic approaches seem to be the effective means for designing multi-criteria grid or cloud schedulers by trading-off various preferences and goals of the system users, resource and service managers and resource owners. The aforementioned technologies are the foundations for the intelligent scalable computing, future generation grid computing and, recently, green computing technologies in grids and clouds.

Although, most of the metaheuristics attempt to find an optimal solution with respect to a specific fixed fitness measure. In the case of evolutionary or genetic algorithms, a great deal of effort has gone into designing efficient representation schemes and genetic operators so as to produce rapid convergence for a good solution. The rapid decrease in diversity of the population results in a highly fitted but homogeneous population, which does not allow the algorithm to perform well in large-scale dynamic environments. Parallelization and hybridization have proven to be an effective solution for increasing diversity in the population and to evolve robust solutions that are able to track the optima. However, the cost of their implementation and execution in the large-scale systems may be very high. Therefore, it is most important to investigate a novel general framework for modeling the mono- and multi-population evolutionary-based metaheuristics to enable the secure access to data and resources, flexible communication, efficient scheduling, self-adaptation, decentralization, and energy-awareness of the system.

This book discusses the advanced research on the effective scalable genetic-based heuristic approaches to grid scheduling, where new scheduling criteria, such as system reliability, security, and energy consumption are introduced and incorporated into a general scheduling model. It serves as a monograph book, which covers the recent hot topics in design, administration and management of the dynamic grid environment with a special emphasis on the preferences and autonomous decisions of the system users, secure access to the processed data and services, and green technologies in computational grids. The book consists of the eight chapters structured into four main parts:

I. *Scheduling Problems in Grid Computing*: collectively known as computational resources or simply infrastructure, computing elements, storage, and services represent a crucial component in the formulation of intelligent decisions of the grid users at all system levels. The first two chapters introduce the general concepts of Computational grids, the types of grid users and scheduling problems in the dynamic grid environment. An ETC Matrix model for Independent Batch Scheduling problem is presented and a number of scheduling constraints, criteria, and scenarios are discussed throughout the text.
II. *Multi-Level Genetic-Based Hierarchical Grid Schedulers*: Chapters 3 and 4 present a concept and the results of the empirical evaluation of a multi-level metaheuristic grid scheduler. The main goal of this method is an effective

hierarchical multi-level exploration of the search space by a family of dependent genetic processes.

III. *Security-Driven Scheduling Model for Computational Grid Using Multi-level Genetic Metaheuristics*: Scheduling and resource allocation in today's Computational Grids arises new requirements and challenges not considered in conventional distributed computing environments. Among these new requirements, task abortion and security become needful criteria for grid schedulers. The former arises due to the dynamics of the grid systems, in which resources are expected to enter and leave the system in an unpredictable way. The later appears crucial mainly due to a multi-domain nature of the grid environment. Chapters 5 and 6 showcase techniques, models and concepts for security awareness in grid scheduling.

IV. *Genetic Solutions to 'Green' Scheduling in Computational Grids*: The efficient resource allocation in grids becomes even more challenging when energy utilization, beyond the classical makespan and user's Quality of Service (QoS), is treated as first-class additional scheduling objective. Chapter 7 presents a brief survey on genetic metaheuristic solutions to green computing. Chapter 8 demonstrates the genetic energy-aware grid schedulers applied to the Dynamic Voltage and Frequency Scaling (DVFS) model for the management of the cumulative power energy utilized by the grid resources.

I believe that this book ought to serve as a reference for students, researchers, and industry practitioners interested or currently working in the evolving interdisciplinary area of intelligent scheduling and resource management models using emergent distributed computing paradigms. I hope that the readers will find new inspiration for their research in high performance computing by seeing the old scheduling problems from a newer and a unique perspective.

Bielsko-Biała, Kraków

Joanna Kołodziej
June 2011 – January 2012

Acknowledgements

It is with great pleasure to express my thanks and gratitude to many people who have been involved in the development of the book and contributed to my projects, publications, and research. Without their support the book could not have been successfully completed.

I am deeply grateful to Prof. Zbigniew Michalewicz for his great expertise, passion, understanding, and inspiration for my research on metaheuristics and evolutionary computation. I would like to convey my special thanks to Prof. Tadeusz Burczyński who was abundantly helpful and offered invaluable assistance, support and guidance. I wish to express my sincere gratitude to Prof. Fatos Xhafa for his support throughout my research on independent grid scheduling and for providing me an opportunity to extend and modernize his grid simulator.

I would like to show my greatest appreciation to Prof. Samee Ullah Khan. I cannot say thank you enough for his tremendous support and help. Without his encouragement and guidance this book and the most of my recent research results would not have materialized.

I am grateful to all my colleagues from the Department of Mathematics and Computer Science, University of Bielsko-Biała and especially the Head of the Department, Prof. Kazimierz Nikodem, for being the surrogate family during the many years of my work there and for their care and attention.

My special thanks go to Prof. Janusz Kacprzyk, Editor-in-Chief of "Studies in Computational Intelligence" Springer Series, Dr. Thomas Ditzinger, Holger Schäpe and all Springer's editorial team for their assistance and excellent cooperative collaboration in this book project. I am also grateful to William Knottenbelt, Philip Moore, Horaco-González-Veléz, Katherine Irvin and Janusz Wojtusiak for their proofreading work.

Finally, I am forever indebted to my family and close friends for their understanding, endless patience and encouragement when it was most required.

Joanna Kołodziej

Contents

Acronyms

Acronym	Definition
ACO	Ant Colony Optimization algorithm;
ANN	Artificial Neural Network;
BC	Branch Comparison operator;
CMOS	Complementary Metal-Oxide Semiconductor;
CPGA	Cellular Parallel Genetic Algorithm;
CPU	Central Processing Unit;
CG	Computational Grid;
CMP	Chip Multiprocessor;
CX	Cycle Crossover;
CVB	Coefficient-of-Variation method;
DAG	Directed Acyclic Graph;
DEM	Dynamic Energy Management;
DFG	Data Flow Graph;
DPS	Dynamic Performance Scaling;

Acronym	Definition
DSP	Digital Signal Processors;
DVFS	Dynamic Voltage and Frequency Scaling;
DVS	Dynamic Voltage Scaling;
ECS	Energy-Conscious Scheduling heuristic;
EC	Evolutionary Computation;
ETC	Expected Time to Compute (matrix);
GA	Genetic Algorithm;
GE-HPGA	Grid-Enabled Hierarchical Parallel Genetic Algorithm;
Grid-sys-admin	Grid System Administrator;
Grid-sp	Grid Service Provider;
Grid-seu	Grid Service End-User;
Grid-pu	Grid Power User;
Grid-pua	Grid Power User Agnostic;
Grid-puds	Grid Power user Developing a Service ;
HGS	Hierarchical Genetic Strategy;
HGS-Sched	Hierarchical Genetic Scheduler;
HPDS	High Performance Distributed System;
IT	Information Technology;
LAN	Local Area Network;
LJFR-SRFR	Longest Job to Fastest Resource - Shortest Job to Fastest Resource;

Acronym	Definition
MAs	Memetic Algorithms;
Max-Min	heuristic;
MB	Meta-Broker;
MCT	Minimum Completion Time;
MDBP	Multi-Dimensional Bin-Packing problem;
MEB	Minimum Energy Broadcast;
MET	Minimum Execution Time;
MI	Millions of Instructions;
MIPS	Millions of Instructions Per Second;
MOGA	Multi-objective Genetic Algorithm;
MOPGA	Multi-objective Parallel Genetic Algorithm;
NSGA	Non-dominated Sorting Genetic Algorithm;
OLB	Opportunistic Load Balancing ;
PC	Personal Computer;
QoS	Quality of Service;
PMX	Partially Matched Crossover;
PSO	Particle Swarm Optimization algorithm;
SA	Simulated Annealing;
SEM	Static Energy Management;
SIMD	Single Instruction Multiple Data;
SO	Sprouting operator;

Acronym	Definition
SOA	Service Oriented Applications;
TS	Tabu Search;
VGrADS	Virtual Grid Application Development Software project;
VLAN	Virtual Local Area Network;
VO	Virtual Organization;
WAN	Wide Area Network;
WSN	Wireless Sensor Networks;

Notation

Notation	Definition
Classification of the Grid Scheduling Problems	
Rm	network of computing resources of various speed
b	batch scheduling mode
im	immediate scheduling mode
dep	dependency among tasks
$indep$	independent scheduling
sta	static scheduling
dyn	dynamic scheduling
C	centralized grid architecture
D	decentralized grid architecture
$H(i)$	hierarchical grid architecture
d_j	deadline constraint for task j

Notation	Definition
sim	multi-objective optimization in simultaneous mode
$hier$	multi-objective optimization in hierarchical mode
C_{max}	makespan
F	flowtime
F_j	time of finishing the task j
$Tasks$	a set of tasks submitted to the grid
$Schedules$	a set of all possible schedules
Lat_{max}	maximum lateness
Lat_j	lateness for the task j
$Tard$	total weighted tardiness
$Tard_j$	tardiness for the task j
$\tilde{w}_j$	weight coordinate for T_j

Independent Batch Scheduling

Notation	Definition
n	the number of tasks in a batch
m	the number of machines available in the system for the execution of a given batch of tasks
$N = \{1, \ldots, n\}$	the set of tasks' labels
$M = \{1, \ldots, m\}$	the set of machines' labels
$objectives$	the set of the scheduling objective functions

Notation	Definition
wl_j	workload of task j
$WL = [wl_1, \ldots, wl_n]$	workload vector
cc_i	computing capacity of machine i
$CC = [cc_1, \ldots, cc_m]$	computing capacity vector
$ready_i$	ready time of machine i
$ready\_times = [ready_1, \ldots, ready_m]$	ready times vector
$ETC[j][i]$	completion time of task j on machine i
$exec_{ave}$	the estimated execution time of all tasks on an 'average' machine in the system
$tvar_{tasks}$	the variance in the execution times of task
$mvar_{mach}$	the variance in the heterogeneity of Grid resources
$\mathscr{S}$	the set of schedules encoded by using the direct method
$\mathscr{S}_{(1)}$	the set of schedules encoded by using the permutation-based method
S	schedule vector in direct representation
Sch	schedule vector in permutation-based representation
$completion[i]$	completion time of machine i
$F[i]$	flowtime for machine i
$Tasks(i)$	the set of tasks assigned to machine i

Hierarchical Genetic Scheduler (HGS-Sched)

Notation	Definition
t	degrees of the the branches
$\widetilde{Max}$	maximal degree of the branches
$1 \geq pop_1 \geq pop_2 \geq \ldots \geq pop_{\widetilde{Max}}$	cardinalities ('sizes') of populations in the branches
$\mu_1 \geq \mu_2 \geq \ldots \geq \mu_{\widetilde{Max}}$	mutation parameters (rates) in the branches
$P^{\bar{e}}_{(r,t)}$	population in the branch of degree t
Met_α	α-periodic metaepoch
SO	sprouting operator
suf	neighborhood parameter
S_t	the length of the suffix of schedule S
f_{hash}	hash function

Security and Game Parameters

Notation	Definition
SD	security demand vector
sd_j	security demand parameter for the task j
TL	trust level vector
tl_i	trust level parameter for the machine i
Pr_f	machine failure probability matrix
$Pr_f[j]i]$	probability of failure of the machine i during the execution of the task j
$completion^s[i]$	the completion time of the machine i in the secure mode

Notation	Definition
$C_{max}(sec)$	makespan in secure mode
$F(sec)$	flowtime in secure mode
$Fail_r$	failure rate parameter
$Play$	number of players (grid users)
Q_a	cost function of the user a
Pl_a	strategy vector of user a
Q	multi-cost game function
Q_{Stac}	multi-cost game function in Stackelberg game
Q_{Fol}	aggregate Followers' multi-cost game function
$Q_a^{(ex)}$	user's task execution cost
$Q_a^{(util)}$	resource utilization cost
$Q_a^{(sec)}$	cost of security-assured allocation of the user tasks

Energy-aware Scheduling Parameters

Notation	Definition
s^i	energetic class of the machine i
Pow_{ji}	capacitive power utilized by the machine i for computing the task j
$Vr_{(i)}$	meta-vector of voltage and frequency levels for machine i
$v_{s_l}(i)$	voltage supply for machine i at the DVFS level s_l

Notation	Definition
$f_{s_l}(i)$	scaling parameter for the frequency of machine i at the DVFS level s_l
$E_{ji}(s_l)$	energy utilized for completing the task j on machine i at the level s_l
E_i	cumulative energy utilized by the machine i for computing all tasks assigned to this machine
$Idle[i]$	idle time of machine i
E_{batch}	total energy consumed for scheduling a batch of tasks
mig	relative amount of the migrating individuals in *IGA*
$deme$	size of the sub-population in *IGA*
m_{deme}	number of migrating individuals in each deme
$Im(E)$	relative energy consumption improvement rate

Part I
Scheduling Problems in Grid Computing

Chapter 1
Scheduling Problems in Hierarchical Grid Environment

Abstract. Scheduling and efficient resource management are the key issues for grid computing. This chapter reveals the complexity of the scheduling problem in Computational Grids under various criteria and users' requirements. This chapter presents a simple taxonomy of *Grid Systems* and defines the hierarchic multi-layer system model along with a general characteristics of the class of grid users, their functions and requirements related to scheduling in grid computing. It introduces a general classification and notation of the grid scheduling problems and the basic scheduling attributes. This chapter ends with a short discussion on the grid users behavior scenarios related to scheduling and a brief survey of the realistic scheduling models.

1.1 Introduction

Grid computing is one of the most popular combinations of traditional distributed computing and utility computing. This combination greatly facilitates today's Information Technology (IT) outsourcing and has become very effective in solving large-scale complex problems from a variety fields, such as social and biological sciences, engineering, and finance. Computational Grids (CGs) are primarily concerned with the development of high-performance applications, which can be executed simultaneously on multiple computers or supercomputers connected by the wide area networks.

Task scheduling and resource allocations are the key issues for CGs. Specifically, in large-scale CGs, distributed resource clusters work within different autonomous domains with their own access policies, which impacts the successful execution of the grid applications across the domain boundaries.

The concept of the 'Grid' has been developed over the past 20 years as part of grid computing project aimed at linking geographically dispersed supercomputers. The grid idea was popularized by the late 1990s by Foster et al. [48] who developed the Globus Toolkit as a general middleware for grid systems [47].

J. Kołodziej: Evolutionary Hierarchical Multi-Criteria Metaheuristics, SCI 419, pp. 3–18.
springerlink.com

Today's *Grid Systems*, are considered along with the *Cluster* and *Cloud* systems, as a principal category of modern High Performance Distributed Systems (HPDSs) as presented in Fig. 1.1.

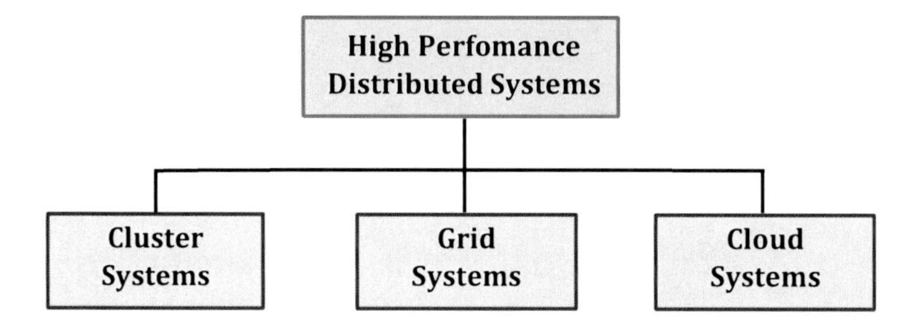

Fig. 1.1 The main categories of High Performance Distributed Systems (HPDSs)

Despite the fact that all HPDS's models are collectively based on the distributed computing paradigm, they have different characteristics which are crucial for the potential system users. Table 1.1 presents a comparison of the main features for three HPDS classes.

Table 1.1 Main features of three HPDS classes

Feature	Cluster	Grid	Cloud
Scale	Small and medium	Large	From Small to Large
Network type	Private LAN	Private, WAN	Public, WAN
Administrative domain	Single	Multi	Both
Resources' domain structure	Homogeneous	Heterogeneous	Heterogeneous
Security	Very High	High	Low

The modern *Cluster Systems* are composed of computers usually restricted to a single switch, or group of interconnected switches, within a single virtual local-area network (VLAN). These systems are designed as platforms for processing the data intensive applications, multi-level system management, and the implementation of the scalable methodologies and techniques. Despite executing computer-intensive applications, today's cluster systems are also used for replicated storage and backup servers, which provide the fault tolerance and reliability for applications.

The main goal of the *Grid Systems* is to connect geographically distributed resources through wide area high speed networks (Internet) in order to minimize the cost of their utilization. In contrast with the cluster and other conventional distributed systems, grids account for the different administrative domains with their own access policies, users' privileges and requirements. The management of the grid resources can be very complex mainly as a result of the high system dynamics and additional users' requirements (e.g., security and low cost scheduling).

The *Cloud Systems* class is the most recently developed HPDS category and is based on the Internet infrastructure. The cloud system is a new supplement and delivery model for IT services, involving over-the-Internet provisions of both physical and virtualized scalable resources. The model is based for the most part on the Virtual Grid Application Development Software project (VGrADS), which is sponsored by The National Institute of Standards and Technology (NIST). The current clouds may be seen as a next step in the evolution of the grid environments. The cloud providers deliver common business applications online, while the software and data are stored on physical servers. From the user's perspective, the cloud model requires minimal management and interactions with IT administrators and resource providers. On the other hand, cloud systems need the complex networking, storage and intelligent system configuration to be self-monitoring and in fact also self-healing system. A self-monitoring of all actions in cloud and the systems states is necessary for the automatic balancing of workloads across the physical network nodes in order to optimize the costs of the system utilization. The self-healing feature means that the system guarantees an automatic service restoration in case of failure of any individual physical software or hardware component of the cloud. However, the efficient solutions for the service and resource management in cloud systems still remain as hot open research problems.

1.2 Grid Types and Multilevel Architecture

Although CG remains the most popular grid environment, today's *Grid Systems* have grown far beyond their original intention. Modern grid infrastructures can benefit various complex applications, e.g., collaborative engineering, data processing and exploration, e-Science applications, by providing many of the preceedingly discussed features. This section presents a simple taxonomy and briefly review the most important types of grids and define a multi-layer architecture of CGs.

1.2.1 Types of Grids

The *Grid Systems* class contains several high performing distribution systems with a varying degree of grid characteristics. There are many grid-like systems developed and deployed for a mixture of purposes, namely computational grid, data/storage grid, campus grid, enterprise grid, global grid, knowledge grid, sensor grid, cluster grid, pc grid, commodity/utility grid, etc. The full list of systems can be found in [48].

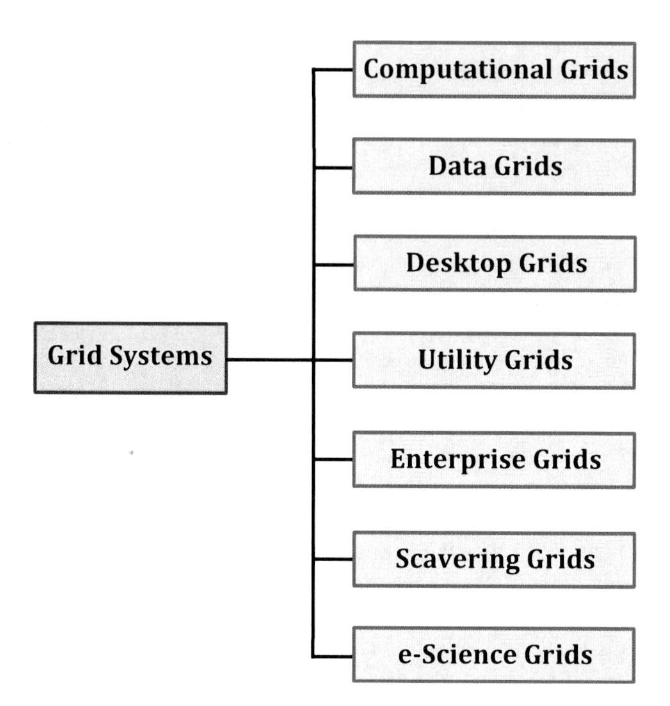

Fig. 1.2 Taxonomy of Grid Systems

A simple taxonomy of the *Grid Systems* is presented in Fig. 1.2, and includes seven of the most prevailing types of grids developed over the past few years.

Computational Grids: CGs were developed for solving complex computational problems that require the processing of large quantities of mathematical operations and/or data. They use the high speed connection network as a, low-cost commodity, medium for accessing the wide spread availability of powerful computing resources. CG systems are a natural extension of the High Performance Computing (HPC) systems, which enable the sharing of a wide variety of computing clusters owned by different geographically distributed organizations [12]. The major CGs' projects include: NASA IPG project [68], the World Wide Grid project [23], the NSF TeraGrid project [144], Nimgrod/G [25] .

Data Grids: Data Grids systems provide the services and infrastructure needed for data-intensive applications to access and modify the large distributed databases resources . All procedures in a Data Grid are mediated by a security layer that handles authentication of entities and ensures the administering of only authorized operations. The main services offered by the system include, but are not limited to, consistency management for replicas, meta-data management, and data filtering [32], [147].

Desktop Grids: The main concept of the Desktop Grids is to connect personal computers (PCs) to large-scale networks by using the Internet or other high-speed wide area networks. To access this feature a PCs owner must simply install an utility application and register in the grid web service. The Desktop Grids architecture is based on the conventional Master-Slave model. The grid applications are split up into many small subtasks that can be processed independently. SZTAKI project [58] is an example of such systems.

Utility Grids: In Utility Grids organizations subscribe to an external utility computing service provider and pay for the utilization of the hardware and software resources. The physical resources in the system are shared and utilized by a number of applications and users from delet numerous organizations. The principal resources offered include, but are not limited to virtual computing environments and storage capacity [50].

Enterprise Grids: Enterprise Grids are made up of grid services and infrastructures, which can be used as support platforms for e-Business and enterprise applications. The Enterprise Grids can process and execute a very wide range of projects; from an investment portfolio risk analysis and pricing securities in the finance and insurance sector; to drug discovery in the pharmaceutical sector; as well as digital media creations. "IBM Grid" project [57], "Oracle Grid" [115] and "HP Grid" project [134] explore some examples of the most widely known Enterprise Grids.

Scavenging Grids: Scavenging Grids are at [33]. Scavenging Grids are considered as an underlying technology of computational and volunteer computing projects, where by the system administrators locate and exploit the Central Processing Unit (CPU) cycles on idle servers and desktop computers in order to compute and execute the users' tasks. Setihome [130] and Rosettahome [125] projects are instances where scavenging grids are used. ***e-Science Grids:*** These systems are defined as the distributed cyber-infrastructures that support scientific investigations performed during global collaborations (particularly those between scientists and their resources). e-Science Grids enable the users to combine and coordinate the research and innovation activities at global levels.

1.2.2 Multi-Level Hierarchical Grid Architecture

The grids are modeled as hierarchical multi-layer systems, which are the result of the hybridization of centralized and decentralized resources and service management. The hierarchy usually consists of two or three levels, depending on the system knowledge, access to data and resources and the organization of the scheduling process. The general concept of multi-level hierarchical grid architecture is presented in Fig. 1.3.

An example of a two-level grid architecture is the Meta-Broker model (MB) [50]. In this system, grid users submit their applications to the MB system (the inter-site level), which uses the information supplied by the resource owners (the intra-site level) to assign user and application requirements to the appropriate user and their machines.

A three-level grid system example is presented by Kwok et al. in [92]. All types of grid users work at global, inter-site and the intra-site levels. At the intra-site level,

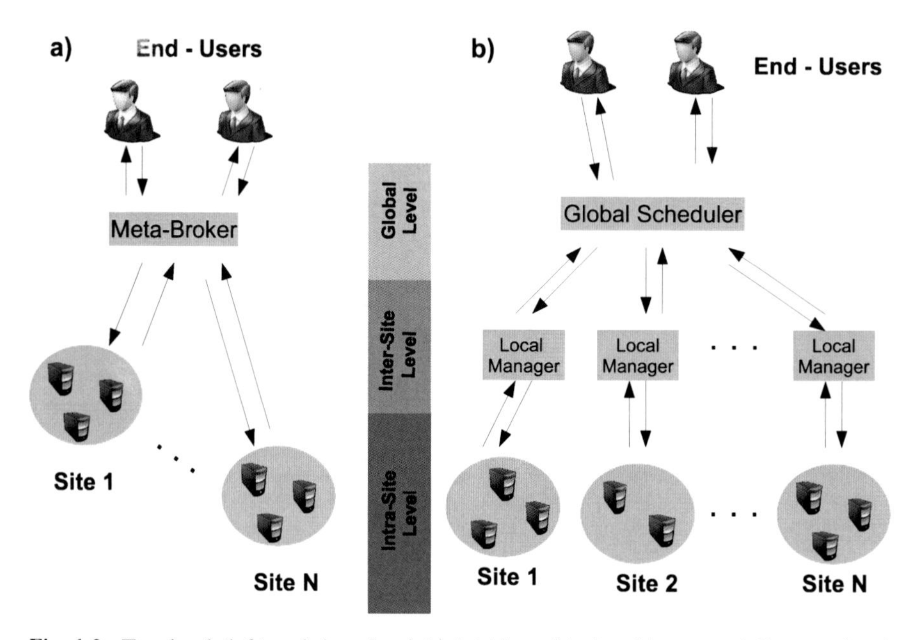

Fig. 1.3 Two-level (left) and three-level (right) hierarchical architecture of Computational Grids

there is an alliance of autonomous machines. The resource owners send information about the computational capacities of the machines to local managers, who define the "grid sites reputation indexes" and forward them to the global scheduler. At the global level the scheduler performs the tasks adaptively, using scheduling algorithms and suggested resources based on availability.

The characteristics of each of the grid levels may be difficult to interpret, mainly because of the intricate and intertwining multi-layer structure of the grid components that span the hierarchical architecture of the system. The four key layers of grid components can be defined by: (1) grid 'fabric' layer, (2) grid core middleware, (3) grid user layer, and (4) applications [27]. The grid 'fabric' layer is composed of the grid resources, services, and local resource management systems. The grid core middleware provides services related to security and access management, remote job submission, storage, and resource information and scheduling. The grid users' layer contains all of the grid service end-users and system entities, and plays the most important role in ensuring multi-criterion scheduling and resource management efficiency.

1.3 Users' Layer in the Grid System

In conventional distributed computing environments just a few categories of users can maintain the whole system. These users' characteristics include operating in the same administrative domains, restricted requirements and actions to small area networks and clusters. In CG systems, due to high heterogeneity of services and resources, as well as hierarchical grid architecture, a large community of the system users co-exist and perform their tasks with different, and sometimes conflicting requirements. These conflicts imply that there are a variety of users' relationships, behavior, and scheduling scenarios. The IT resources generally belong to different owners (institutions, enterprises, or individuals) and are managed by different administrators. Every resource administrators can potentially adopt different sets of rules and configuration directives, as well as impose different usage policies on the system users. In addition, the task managers, resource providers, and local schedulers may each play a different role if the new conceptual scheduling criteria such as energy awareness, security, and resource reliability are considered. Therefore there is a need to analyze and model such users' requirements and relations to predict the users' actions, and to optimize the schedulers' behavior as well as the whole system performance at both the individual and global levels.

1.3.1 Main Types of Grid Users

In large-scale dynamic grid environments the users play an assortment of roles at the different system's levels, these include working as the: grid administrators, community or Virtual Organization (VO) administrators, node administrators, service owners, users' groups administrators, service end-users, etc. Based on the grid users taxonomy defined in [111], below this chapter will present a brief profile of six main types grid users, namely Grid System Administrators, Grid Service Providers, Grid Service End-Users and three categories of Grid Power Users [88], [89].

Grid-sys-admin: *Grid System Administrator*. The main function of system administrator is the management of grid nodes and clusters security, infrastructure delivery, and system configuration. The system administrator is an expert in computer science and is also in charge of monitoring the users' activities and system performance.

Grid-sp: *Grid Service Provider*. Users in this role also have expertise in computer science, as well as authorization and personalized identity management. This role provides authentication, authorization, and accounting for the Grid Service-End users.

Grid-seu: *Grid Service End-User*. The Grid Service End-User's main focus is the submission of task and applications to the grid schedulers and managers. The role also requires the ability to specify time and budget constraints for the task scheduling and execution systems. Additionally, individuals in this role upload the data, scripts or source codes necessary for solving the submitted tasks and the execution of applications, as well as run queries, executable code or scripts with the assistance of the grid Service Provider . Unlike many of the other roles the Grid-seu does not need to be an expert in computer science or with the grid architecture and management.

Grid-pua: *Grid Power User Agnostic* of grid resource node. Individuals in this role focus mainly on activities associated with program development and data management. The Grid-pua is not usually concerned with where the processing takes place in the grid system.

Grid-pu: *Grid Power User* requiring specific grid resource nodes. The main duties of this role consist of activities similar to those of the Grid-pua's. In addition, the grid service and resource owners may wish to have a direct authentication, authorization, and accounting relationship with the Grid-pus (differently from Grid-pua users).

Grid-puds: *Grid Power user Developing a Service* are generally individuals withexpertise in computer science (like Grid-pua and Grid-pus) and service administration and development (similarly to Grid-sp). This role at times may interact with Grid-seu in an tentative manner.

There are a variety of ways the access management requirements and the special polices pertaining to each user group, and defined by the resource owners, may be fulfilled. Each and every user group, should adapt in response to the requirements of the users' communities. Table 1.2 presents the users' general requirements with respect to the grid administration, service, application and porting, usability and resource utilization (see also [40]).

An important concept to address, as it is imperative across all grid levels, is the standardized authentication and authorization mechanisms as well as the globally accepted trustworthiness of the grid user. The trustworthiness of the grid user can be defined as a user authentication trustworthiness parameter (UAT), which is expressed as a degree of the user's approved system authentications, and must be taken as a basic qualification in judging access requests. The system makes an access control decision based on the user's authentication trustworthiness. Although the user has passed system authentication, one cannot be certain whether that individual should be trusted or not. There are some uncertainties in authentication systems, i.e., uncertainties of the authentication mechanisms, authentication rules, and authentication conclusions. These uncertainties in the authentication process can be modeled using the Fuzzy Logic. The Fuzzy Logic has been widely accepted in grid computing for intrusion detection systems purposes, as well as for the prediction of the users' actions and decisions [3, 167]).

The problems of authentication trustworthiness have been extensively studied, analyzed, and applied in a significant volume of publications in the domain [150]. A majority of the methods are based on the concepts similar to those of the Virtual Organization (VO) model. User's who acquire membership with the VO are authorized and granted access to VO's resources. The VO membership contributes to the distribution of the user's management overheads as well as reduces the procedures in replicating administrative efforts across the grid. This concept of VOs is also a key issue in the grid architecture development. The proper designing of a CG architectural model is one of the most important concerns in ensuring efficient resources, as well as task and user management according to the various users' requirements. The hierarchical layered structure of the system in conjunction with user relations is sufficient to capture the realistic administrative features of a real-life, large-scale distributed CG environment.

Table 1.2 Grid Users' general requirements

Requirement type	User type
Grid General Requirements	
Reliable grid middleware; Quality of Service (QoS)	All users
Fine grained simple access policies to data and databases	All users
Monitoring the grid jobs and tasks, estimation of task queue delay	All users
Standardized authentication and authorization mechanisms	All users
Globally accepted, trustworthy grid user identity	All users
Administration Requirements	
Secure communication and transfer	Grid-sys, Grid-puds Grid-seu
Uniform configuration across all grid components	Gris-sys-admin, Grid-puds
Upgrading and the management of the software	Grid-pua, Grid-pu Grid-sys-admin
Testing and monitoring of grid components, service recovery, outages and maintenance scheduling	Grid-sys-admin
Service Requirements	
Encryption and protection of data on grid storage elements	Grid-sp, Grid-pua Grid-pu
Fast access and reliable transfer of massive amounts of data	Grid-sp, Grid-seu Grid-sys-admin
Ad-hoc integration of external arbitrary data sources	Grid-sp
Application and Porting Requirements	
Consistent API for all middleware components	Grid-sys-admin
Standardized error codes and error handling procedures	All users
Utilization and Usability Requirements	
Visualization of the computational results	Grid-seu
Safe and easy authentication procedures	Grid-seu, Grid-sp
Reliable real-time and instantaneous task submission for high priority tasks for e.g. risk and disaster management, recovery, etc.	Grid-seu

1.3.2 Grid End Users' Requirements for Scheduling

The complexity of the scheduling scenarios and the roles of the system administrators, service providers, and grid power users strongly depend on the Grid-seu's QoS requirements and specific scheduling and task execution criteria, which is defined as follows:

Requirements for Specifying a Single Computational Task. At the basic level, the Grid-seu needs to be able to specify and submit a single monolithic application (with well defined input and output data), a bag-of-tasks application with no dependencies among them, and complex parallel applications. The user may also be required by the system to provide information on types of tasks (e.g. data intensive *vs.* CPU intensive computing) and an estimation of task workload. In many cases, the users should be able to submit their tasks/applications as either executable or source code, which need to be compiled and linked for further execution. Some source codes may require a software deployment not available in the grid cluster. Due to the complexity of the compilation and execution processes on heterogeneousarchitectures within the grid, it is recommended to build and test applications on a specific platform prior to submitting them to the system. Finally, in most cases data is assumed to be shipped with the task/application.

Requirements for Specifying a Job of Multiple Tasks. The user should be able to specify a complex job involving the execution of multiple tasks with internal relations among all the components. The input and output data must be defined for each task, providing the ability to specify relational data. Some graphical interfaces and representation of the graph structures (like Directed Acyclic Graph (DAG)) may also be provided for the specification of tasks inter-relations.

Access to Remote Data. The input and output data specified by the user may be stored remotely. Therefore, users will need to provide the location of the remote data. If a ubiquitous wide-area file system is in operation on the grid, the user would only have to care about the location of files and data with respect to some root location under which they are stored.

Resource Specification. The user may specify special requirements for the resources necessary in optimizing the execution times and costs of scheduling and computing tasks. The user may wish to target particular types of resources (e.g. SMP machines), but should not be concerned with the type of resource management on the grid, nor with the resource management systems on individual resources on the grid.

Resource reliability. In some cases, the machines within the grid system could be unavailable due to high system dynamics or special policies of the resource owners. The user should be informed about the resource reliability in order to reduce the cost of possible resource failures or the abortion of executed tasks. In the case of resource failure the system administrators (Grid-sys-admin users) can activate re-scheduling or task migration procedures, and preemption policies.

Trustfulness of Resources - Secure Scheduling. The user may be required to allocate his tasks in the most trustful resources. Therefore the user should be able to verify the trust indexes of the resources and estimate the security demands for his tasks on the available resources.

Standardized authentication and authorization mechanisms requirements. The CG's users will likely utilize a standardized certificate authentication scheme. The certificates can be digitally signed by a certificate authority, and kept in the user's repository, which is recognized by the resources and resource owners. It is desirable for a certificate to be automatically created by the user's interface application during task submission.

Job Monitoring and Control. The grid users should be able to monitor the current status of their tasks and applications in the system.

1.4 Scheduling Attributes and Problem Types

The main aim of scheduling in large-scale distributed computational environments is efficient mapping of tasks originated by applications to a set of available resources. The tasks and resources can be added and dropped to and from the system. Scheduling in computational grids remains a challenging NP-complete global optimization problem due to the heterogeneous structure of the system and co-existence of local geographically dispersed job dispatchers and resource owners working in different autonomous administrative domains.

This section defines the main scheduling attributes and introduces a general classification of the scheduling problems in CGs and the scheduling criteria.

1.4.1 Scheduling Attributes

Different types of scheduling problems in CGs may be defined with respect to different properties of the underlying grid environment and various requirements of the users. To achieve the desired performance of the system, both users' conditions and grid environment information must be "embedded" into the scheduling mechanism [2], [157], [86].

Fig. 1.4 depicts four main scheduling attributes that must be setup to specify a particular tasks-machines mapping problem, namely: (a) the environment, (b) grid architecture, (c) task processing policy, and (d) tasks' interrelations.

The general scheduling scenario in CGs may be realized in static or dynamic environments. In the case of the static scenario the number of the submitted applications and the available resources remain constant in a considered time interval, while in the dynamic scenario the resources may be added or removed from the system in an unpredictable way.

The resource management and scheduling can be organized in centralized, decentralized, or hierarchical modes. In *centralized model*, there is a central authority, who has a full knowledge of the system. The primary disadvantages of this model is its

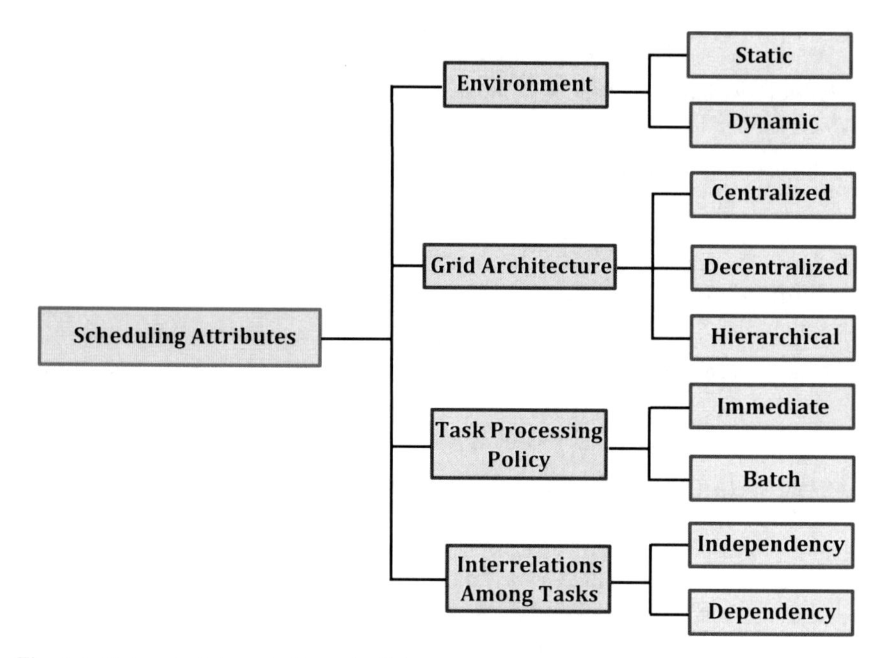

Fig. 1.4 Main scheduling attributes in CGs

limited scalability, lack of fault tolerance, and the difficulty in accommodating multiple local policies imposed by the resource owners. In *decentralized model*, local schedulers interact with each other to manage the tasks pool. In this model, there is no central authority responsible for resource allocation. Hence, the model naturally addresses issues such as fault-tolerance, scalability, site-autonomy, and multi-policy scheduling. Finally, in the *hierarchical model*, there is a central meta-scheduler (or meta-broker), which interacts with local job dispatchers in order to define the optimal schedules. The higher level scheduler manages large sets of resources while the lower level job managers control small set of resources. The local schedulers have knowledge about resource clusters, but they cannot monitor the whole system. The advantage of using hierarchical scheduling is that it incorporates scalability and fault-tolerance issues while also retaining some of the advantages of the centralized scheme such as co-allocation.

A specification of the tasks' processing policy is important in the identification of the particular scheduling problem. In the instantaneous mode the tasks are scheduled as soon as they are entered into the system. In batch scheduling, the submitted tasks are grouped into batches and the scheduler assigns each batch to the resources.

Finally, tasks may be independently submitted and calculated in the grid system or considered as parallel applications with priority constraints and interrelations among the application components (usually modeled by a Directed Acyclic Graph (DAG)).

1.4.2 Grid Scheduling Notation and Criteria

Research conducted produced no known standard notation for classification of the scheduling problems in CGs. Fibich et al. [45] proposed an extension of the Graham's [56] and Brucker's [21] classifications of scheduling problems. The characteristics of the resource-constrained project scheduling problem [22] and resource-constrained machine scheduling [17] [16] might be helpful in specifying and formal describing the grid resources.

Based on the methodology presented in [45] and [78] and the main scheduling attributes specified in the previous section, the basic notation for the grid scheduling problem instances in CGs can be defined as follows:

$$\alpha|\beta|\gamma \tag{1.1}$$

where α characterizes the resource layer and grid architecture type, β specifies the processing characteristics and the constraints, and γ denotes the scheduling criteria.

(α) Resource Characteristics and Grid Architecture Type

According to the standard resource notation [56] the grid computational resources can be classify as Rm machines[1] with possible different speeds for different jobs. grid architecture type can be denoted by C for centralized, D for decentralized and $H(i)$ for hierarchical system, where i denotes the system levels. The following notation

$$Rm, H(3) \tag{1.2}$$

is used for the representation of the 3-level hierarchical grid with heterogeneous resources. The detailed characteristics of each resource may be specified by using special local policies and characteristics according to the users' and applications requirements [17] [16].

(β) Tasks Processing Mode, Tasks Interrelations, Static and Dynamic Scheduling Modes and Scheduling Constraints

The following notation can be used for setting the grid tasks' attributes and scheduling modes:

- b – batch mode;
- im – immediate mode;
- dep – dependency among tasks ;
- $indep$ – independent scheduling;
- sta – static scheduling;
- dyn – dynamic scheduling;

[1] For single CPU machine α=1; identical machines in parallel infrastructure – Pm, machines in parallel with different speeds – Qm, unrelated machines in parallel –Rm, and (flow—open—job) shop resources (Fm, Om, Jm).

The scheduling may be conducted under various constraints specified by all the grid users. The typical constraints include budget and deadline (d_j) limits for a given task j. An instant of the independent batch scheduling in dynamic mode with a limited deadline can be denoted as follows:

$$b, indep, dyn, d_j. \tag{1.3}$$

(γ) Scheduling Criteria and Objectives

The problem of scheduling tasks in CG is multi-objective in its general setting as the quality of the solutions can be measured using several criteria.

Two basic models are utilized in multi-objective optimization: hierarchical and simultaneous modes. In the *simultaneous mode* (*sim*) all objectives are optimized simultaneously while in the *hierarchical* (*hier*) case, the objectives are sorted *a priori* according to their importance in the model. The process starts by optimizing the most important objective and when further improvements are impossible, the second objective is optimized under the restriction of keeping unchanged (or improving) the optimal values of the first, and so on. It is very hard in grid scheduling to define or efficiently approximate the Pareto front, especially in dynamic scheduling[2]. This set of Pareto optimal solutions may extend very fast together with the scale of the grid and the number of the submitted tasks. The specification of the structure of the Pareto front is also an open problem in grid scheduling. Due to the sheer size of the grid itself and huge number of possible schedulers event for a small amount of tasks, an effective and fast exploration of the search space for the problem is very difficult. The knowledge of the optimization landscape is usually limited just to the small-area clusters [41].

The main scheduling criteria can be divided into two classes: grid system performance criteria and optimization criteria [157]. Grid system performance criteria include CPU utilization of grid resources, load balancing, system usage, queuing time, throughput, turnaround time, cumulative thorough output. Recently this class was extended by the resource failure rates and energy consumption, which will be discussed in details in Chapters 6-7.

The main scheduling optimization criteria include: makespan, flowtime, resource utilization, load balancing, matching proximity, turnaround time, total weighted completion time, lateness, weighted number of tardy jobs, weighted response time, etc.

Four basic scheduling objectives for grid scheduling are:

- The *makespan* is defined as the finishing time of the latest task and can be calculated by the following formula:

$$C_{max} = \min_{S \in Schedules} \left\{ \max_{j \in Tasks} C_j \right\}, \tag{1.4}$$

[2] A solution is Pareto optimal if it is not possible to improve a given objective function without deteriorating at least another one [142].

where C_j denotes the time when task j is finalized, *Tasks* denotes the set of all tasks submitted to the grid system and *Schedules* is the set of all possible schedules.

- The *flowtime* is expressed as the sum of finalization times of all the tasks. It can be defined in the following way:

$$F = \min_{S \in Schedules} \left\{ \sum_{j \in Tasks} C_j \right\} \tag{1.5}$$

- The *maximum lateness* is calculated as follows:

$$Lat_{max} = \max_{j \in Tasks} Lat_j, \tag{1.6}$$

where L_j denotes the lateness for the task j and

$$Lat_j = C_j - d_j \tag{1.7}$$

where d_j is the deadline for task j.

- A *total weighted tardiness* is calculated by using the following formula:

$$Tard = \sum_{j \in Tasks} w_j Tard_j \tag{1.8}$$

where w_j is a weight coordinate and $Tard_j = max(Lat_j, 0)$.

The detailed definition of the rest of the optimization metrics can be found for example in [157]. The class has been recently extended by the security criteria [87].

The notation

$$hier, (C_{max}, F) \tag{1.9}$$

means that makespan and flowtime are optimized in the hierarchical mode.

Both performance and optimization criteria are desirable for any grid system; however, their achievement depends on the considered model (batch system, interactive system, etc.). It should be stressed that these criteria can be conflicting; for instance, minimizing makespan conflicts with resource usage and response time.

1.5 Summary

The main aim of this chapter was the overall characteristics of the hierarchical dynamic CG system, its components and users. Modern grid computing systems, which are made up of hundreds or thousands of various components (computers, databases, etc) must provide (in fact) a whole range of services and not just a high performance computing platform. Various types of information and data processed in today's CGs may be incomplete, imprecise, fragmentary and overloading, which complicates the specification of proper evaluation criteria, assignment scores, availability of resources, and the final collective decisions of the users. The system complexity may also be the reason for higher energy consumption. It makes that the design of the efficient grid resource management model remains the challenging problem for the researchers and practitioners. Scheduling problems in grids are be considered as a family of NP-complete optimization problems. Depending on the restrictions imposed by the application needs, the complexity of the problem can be determined by the number of objectives to be optimized, such as (single vs. multi-objective), the type of the environment (static vs. dynamic), the processing mode (immediate vs. batch), and tasks interrelations (independence vs. dependency). This chapter presented a novel categorization of the grid scheduling problems. This classification may be easily extended if the detailed specification of tasks and system resources is provided.

Chapter 2
Independent Batch Scheduling: ETC Matrix Model and Grid Simulator

Abstract. This chapter addresses the problem of *Independent Batch Scheduling* in Computational Grids (CGs). The Expected Time to Compute (ETC) matrix model is defined and employed for the specification of the main scheduling objectives, namely makespan and flowtime, in terms of completion times of the grid computational nodes. This chapter ends with a outline of the main concept of the grid simulator dedicated to the batch scheduling. This simulator is used in the experimental analysis presented in the rest of this book.

2.1 Introduction

Independent Batch Scheduling is a fundamental model of scheduling in grid systems. In this model the tasks are grouped into batches and can be executed independently in a hierarchically structured static or dynamic grid environments. Due to the massive capacity of parallel computation in CGs, this kind of scheduling is very useful in illustrating large amount of realistic scenarios. Real life examples of batch scheduling include: (a) processing of large log data files of online systems (e.g. banking systems, virtual campuses, and health systems), (b) processing of large data sets from scientific experimental simulations (e.g. High Energy Physics and Parameter Sweep Applications), and (c) data mining in bio-informatics applications.

According to the notation introduced in Sec. 1.4.2 an instance of the independent batch grid scheduling problem can be defined as follows:

$$Rm\left[\{b, indep, (stat, dyn), hier\}\right](objectives)) \tag{2.1}$$

where:

- Rm – Graham's notation references that tasks are mapped into (parallel) resources of various speed[1]
- b – designates that the task processing mode is 'batch mode'

[1] In independent grid scheduling it is usually assumed that each task may be assigned just to one machine.

J. Kołodziej: Evolutionary Hierarchical Multi-Criteria Metaheuristics, SCI 419, pp. 19–30.
springerlink.com

- $indep$ – denotes 'independency' as the task interrelation
- (sta, dyn) – indicates that we will consider both static and dynamics grid scheduling modes
- $hier$ – references that the scheduling objectives are optimized in hierarchical mode
- $objectives$ – denotes the set of the considered scheduling objective functions.

2.2 Expected Time to Compute (ETC) Matrix Model

In this section, the following notation for tasks and machines in independent grid scheduling is introduced from this point forward will be used throughout the book:

- n – the number of tasks in a batch;
- m – the number of machines available in the system for the execution of a given batch of tasks;
- $N = \{1, \ldots, n\}$ – the set of tasks' labels;
- $M = \{1, \ldots, m\}$ – the set of machines' labels.

Tasks and machines are characterized by the following parameters:

(a)**Task** j:

- wl_j – workload parameter expressed in Millions of Instructions (MI)
- $WL = [wl_1, \ldots, wl_n]$ is a *workload vector* for all tasks in the batch;

(b)**Machine** i:

- cc_i – computing capacity parameter expressed in Millions of Instructions Per Second (MIPS) , this parameter is a coordinate of a *computing capacity vector*, which is denoted by $CC = [cc_1, \ldots, cc_m]$;
- $ready_i$ – ready time of i, which expresses the time needed for the reloading of the machine i after finishing the last assigned task, a *ready times vector* for all machines is denoted by
  $ready\_times = [ready_1, \ldots, ready_m]$.

Tasks in this model may be considered as monolithic applications or meta-task with no dependencies among the components. The workloads of tasks can be estimated based on specifications provided by the users, or on historical data, or can be obtained from system predictions [64]. The term 'machine' is related to a single or multiprocessor computing unit or even to a local small-area network.

For each pair (j,i) of task-machine labels, the coordinates of WL and CC vectors are used for an approximation of the completion time of the task j on machine i. This completion time is denoted by $ETC[j][i]$ and can be calculated in the following way:

$$ETC[j][i] = \frac{wl_j}{cc_i}. \tag{2.2}$$

All $ETC[j][i]$ parameters are defined as the elements of an ETC matrix , $ETC = [ETC[j][i]]_{n\times m}$, which is the main structure in ETC model. The elements in the rows of the ETC matrix define the estimated completion times of a given task on different machines, and elements in the column of the matrix are interpreted as approximate times of the completion of different tasks on a given machine.

The ETC matrix model can be characterized by three main parameters:

- heterogeneity of resource;
- heterogeneity of task;
- consistency.

Heterogeneity of machine is defined as a variation of the values in rows of the ETC matrix. It is interpreted as a degree of variation of the machine execution times for a given task. *Heterogeneity of task* is defined as a variation of the values in matrix columns. It is interpreted as a degree of variation of the task execution times for a given machine. The averaged values of all tasks and machine heterogeneity parameters define the heterogeneities of tasks and resources in the whole system.

Another feature of the ETC matrix is its consistency . An ETC matrix is *consistent* if for each pair of the machines i and $\hat{i}$ the following condition is satisfied: if the completion time of some task j is shorter on machine i than on machine $\hat{i}$ then all tasks can be executed (and finalized) faster on i than on $\hat{i}$. The inconsistency of the matrix ETC means that there no consistency relation among machines. Semi-consistent ETC matrices are inconsistent matrices having a consistent sub-matrix.

There are numerous methods of generating the ETC matrices, which reflect the machine and task heterogeneity. In the range-based method [100] the heterogeneity of a set of completion times is quantified by the range of the values of those times. Two range parameters are defined as task and machine heterogeneities, and an uniform distribution is used for generating the ETC matrix elements.

In a *Coefficient-of-Variation (CVB)* [6] method the ETC matrix is generated by gamma distributions [93]. The key parameters for this method are defined as follows:

- the estimated execution time of all tasks on an 'average' machine in the system, $exec_{ave}$,
- the variance in the execution times of task, $tvar_{tasks}$,
- the variance in the heterogeneity of grid resources, $mvar_{mach}$.

The parameters $exec_j$ and $tvar_{tasks}$ are used for estimating the execution times $ETC[j][i]$ of the tasks on the machine i with the 'average' speed in the systems. The times $ETC[j][i]$ are generated by using the gamma distribution with the shape and scale parameters denoted by α_t and β_t respectively. That is:

$$ETC[j][i] = Gamma(\alpha_t, \beta_t), \tag{2.3}$$

where:

$$\alpha_t = \frac{1}{tvar_{tasks}^2} \quad (2.4)$$

$$\beta_t = \frac{exec_{ave}}{\alpha_t} \quad (2.5)$$

A vector of $ETC[j][i]$ parameters ($j \in N$) defines one column (indexed by i) of the ETC matrix. Each element of this column is then used for generating one row of the ETC matrix, that is:

$$ETC[j][\hat{i}] = Gamma(\alpha_m, \beta_m[j]), \quad (2.6)$$

where:

$$l\alpha_m = \frac{1}{mvar_{mach}^2} \quad (2.7)$$

$$\beta_t = \frac{ETC[j][i]}{\alpha_m} \quad (2.8)$$

and $\hat{i} \in M$, $\hat{i} \neq i$.

Finally, the Eq.(2.2) may be used for calculating the ETC matrix. The coordinates of WL and CC vectors are generated by using the Gaussian distributions the parameters of which express the heterogeneities of tasks and resources in the system.

2.2.1 Schedule Representation

Schedules in grid computing can be represented by the vectors of machines' or tasks' labels. Two different encoding methods of schedules in grids can be defined in the following way.

Definition 2.1. *Let us denote by $\mathscr{S}$ the set of all permutations* ***with repetition*** *of the length n over the set of machine labels M. An element $S \in \mathscr{S}$ is termed a* schedule *and it is encoded by the following vector:*

$$S = [i_1, \ldots, i_n]^T, \quad (2.9)$$

where $i_j \in M$ denotes the number of the machine on which the task labeled by j is executed.

This encoding method is called *direct representation* of the schedule.

The $\mathscr{S}$ set can be also defined as the Cartesian product of n copies of the M sets. That is to say:

$$Schedules = \underbrace{M \times \ldots \times M}_{n}. \quad (2.10)$$

The cardinality of $\mathscr{S}$ is m^n.

Remark 2.1. In some approaches the $\mathscr{S}$ set may be considered as a discrete subset of an n-dimensional metric space $\mathbb{R}^n$ with the conventional *Euclidean Metrics* restricted to $\mathscr{S}$. The distance of any two schedules $S^1, S^2 \in Schedules$ is calculated by using the following formula:

$$Dist_e(S^1, S^2) = \sqrt{\sum_{j=1}^{n} (S^1[j] - S^2[j])^2} \tag{2.11}$$

The metrics $Dist_e(S^1, S^2)$ can be further normalized and used for the definition of the *similarity* relation for the schedules (see Chapter 3, Sec. 3.2 and 3.3).

The direct representation of the schedules can be easily transformed into a *permutation-based representation* , which is defined as a vector u of labels of tasks assigned to the machines. For each machine the labels of the tasks assigned to this machine are sorted in ascending order by the completion times of the tasks. Formally, this kind of schedule encoding method can be defined in the following way:

Definition 2.2. *Let us denote by $\mathscr{S}_{(1)}$ the set of all permutations* ***without repetitions*** *of the length n over the set of task labels N. A permutation $Sch \in \mathscr{S}_{(1)}$ is called a* permutation-based representation *of a schedule in CG and can be defined by the following vector:*

$$Sch = [Sch_1, \ldots, Sch_n]^T, \tag{2.12}$$

where $Sch_i \in N$, $i = 1, \ldots, n$. The cardinality of $\mathscr{S}_{(1)}$ is $n!$.

In this representation some additional information about the numbers of tasks assigned to each machine is required. The total total number of tasks assigned to a machine i is denoted by $\widetilde{Sch}_i$ and is interpreted as the i-th coordinate of an assignment vector $\widetilde{Sch} = [\widetilde{Sch}_1, \ldots, \widetilde{Sch}_m]^T$, which defines in fact the loads of grid machines.

Example 2.1. The following vector $S = [1,2,1,4,3,1,2,4,3,3]^T$ is an example of the schedule for 4 machines and 10 tasks encoded by the direct representation method. The same schedule in the permutation-based representation is as follows:

$(Sch = [1,3,6,2,7,5,9,10,4,8]^T; \widetilde{Sch} = [3,2,3,2]^T)$.

2.2.2 Scheduling Criteria

ETC matrix model is very useful for the formal definition of all main scheduling objective functions (see Chapter 1, Sec. 1.4.2). The makespan and flowtime may be additionally expressed in terms of the completion times of the machines . A *completion time C_i* of the machine i is defined as the sum of the ready time parameters for this machine and a cumulative execution time of all tasks actually assigned to this machine. The completion time of the machine i is denoted by $completion[i]$ and it is calculated in the following way:

$$completion[i] = ready_i + \sum_{j \in Task(i)} ETC[j][i], \tag{2.13}$$

where $Task(i)$ is the set of tasks assigned to the machine i.

The $completion[i]$ parameters are the coordinates of the following completion vector:

$$completion = [completion[1], \ldots, completion[m]]^T \tag{2.14}$$

Vector C is used for calculating the makespan C_{max}[2] in the following way:

$$C_{max} = \max_{i \in M} completion[i]. \tag{2.15}$$

In terms of ETC matrix model, a flowtime for a machine i can be calculated as a workflow of the sequence of tasks on a given machine i, that is to say:

$$F[i] = ready_i + \sum_{j \in Sorted[i]} ETC[j][i] \tag{2.16}$$

where $Sorted[i]$ denotes a set tasks assigned to the machine i sorted in ascending order by the corresponding ETC values.

The cumulative flowtime in the whole system is defined as the sum of $F[i]$ parameters, that is:

$$F = \sum_{i \in M} F[i] \tag{2.17}$$

A comprehensive list of the scheduling criteria defined in terms of completion times and by using the ETC matrix model can be found in [157].

2.3 Main Concept of the Grid Simulator: *Sim-G-Batch*

Simulation seems to be the most effective method for a comprehensive analysis of the scheduling algorithms in large-scale distributed dynamic systems, such as grid or cloud environments. It simplifies the study of schedulers performances and avoids the overhead of coordination of the resources, which usually happens in the real-life grid or cloud scenarios. Simulation is also effective in working with difficult and highly parametrized problems. In such cases a considerable number of independent runs is needed to ensure significant statistical results. Using the simulators for the evaluation of grid schedulers is feasible, mainly because of the high complexity of the grid environment.

Using the simulators for the evaluation of the grid schedulers is feasible, mainly because of high complexity of the grid environment. Many simulation packages, useful in the design and analysis of scheduling algorithms in grid systems, have been recently proposed in literature. MicroGrid [135] , ChicSim [121] and GridSim [27] are currently the major projects in grid simulation.

---

[2] The notation for the scheduling objectives is the same as in Sec. 1.4.2.

This section presents the main concept of a *Sim-G-Batch* grid simulator for independent batch scheduling,as an extension and modification of the *HyperSim-G* framework [163].

2.3.1 Basic Concept of Sim-G-Batch

Sim-G-Batch is based on the discrete event-based model, which facilitates the evaluation of different scheduling heuristics under a variety of scheduling criteria across several grid scenarios. These scenarios are defined by the configuration of security conditions for scheduling and the access to the grid resources, grid size, energy utilization parameters, and system dynamics. The simulator allows the flexible activation or deactivation of all of the scheduling criteria and modules, as well as works with a mixture of meta-heuristic schedulers. The simulation results and traces are graphically represented and may be saved as files of different formats such as spreadsheets or pdf files. The simulator structure allows for an easy association with the external or internal embedded database systems particularly for storing historical executions.

The main concept of the *Sim-G-Batch* simulator is presented in Fig. 2.1.

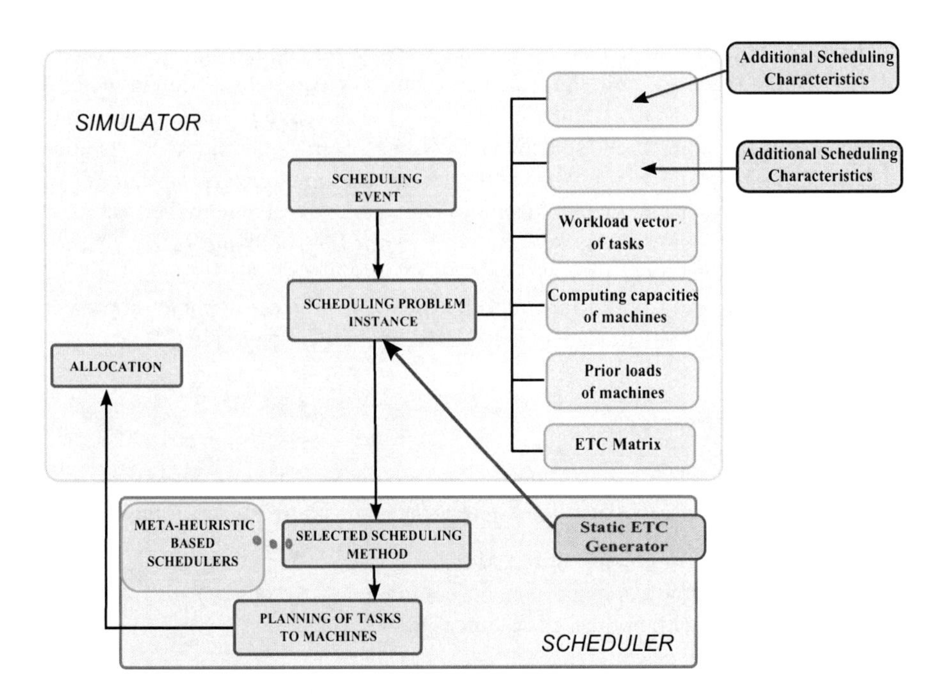

Fig. 2.1 General flowchart of the *Sim-G-Batch* simulator linked to independent batch scheduling in CGs

The *Sim-G-Batch* simulator generates an instance of the scheduling problem by using the following input data:

- the workload vector of tasks ,
- the computing capacity vector of machines,
- the vector of prior loads of machines, and
- the *ETC* matrix of estimated execution times of tasks on machines.

It is important to address that fact that input data may be extended if additional scheduling criteria are or need to be considered (see Chapter 5 and Chapter 8). The static benchmarks for the small grids are generated by using the external *Static ETC Generator* module.

The users can specify their own scheduling scenario by changing the number of tasks and machines. The capacity of the resources and the workload of tasks are randomly generated by using the Gaussian distribution [101]. It is also assumed that all tasks submitted to the system must be scheduled and all machines in the system can be used.

The structure of the *Sim-G-Batch* application is based on the 2-module *HyperSim-G* architecture and it is composed of *Simulator* and *Scheduler* modules. The main simulation flow can be defined as follows. When a scheduling event is triggered, the *Simulator* creates an instance of the scheduling problem, based on the current task batch and the pool of available machines. The *Simulator* computes an instance of the scheduling and passes it on to the *Scheduler*, which activates a scheduling method specified by the user of the simulator software. The *Scheduler* generates the optimal schedules according to the specified scheduling criteria and sends the schedules back to the *Simulator*. The *Simulator* makes the allocation of the grid resources and re-schedules any tasks assigned to machines which are unavailable in the system.

The *Sim-G-Batch* software was written in C++ for Linux Ubuntu *10.10*. The access to selected modules and resolution methods is available through the Web service, which is the result of work on the WebGridUPC project – a common project with Technical University of Catalonia in Barcelona (UPC Spain) [151][3].

2.3.2 Key Parameters

The simulator is highly parameterized in order to illustrate the typical realistic grid scenarios. The main parameters of *Sim-G-Batch* can be interpreted as follows:

- *Init. hosts*: Number of hosts initialized in the grid environment.
- *Max. hosts*: Maximum number of resources in the grid system.
- *Min. hosts*: Minimum number of resources in the grid system.
- *MIPS*: Computing capacity of resource.
- *Add host*: The frequency of activation of the new resources in the system.
- *Delete host*: The frequency of deactivation of the idle or failed resources in the system.

[3] The codes of the benchmars and meta-heuristics are available upon request to Fatos Xhafa (www.lsi.upc.edu/fatos) or Joanna Kołodziej (www.joannakolodziej.org)

- *Total tasks*: Total number of tasks in the batch.
- *Init. tasks*: Initial number of tasks in the system .
- *Workload*: Workload of task.
- *Interarrival*: Frequency of submission of new tasks to the system .
- *Reschedule*: Re-scheduling policy.
- *Number runs*: Number of independent runs of the simulator with the same configuration of the parameters.
- *Scheduling strategy* : The type of the scheduler, maximal execution time of the scheduler (in seconds), and the optimization mode of the scheduling objective functions.

The initial number of machines in the system is defined by the parameter *Init. number of hosts*. The parameters *Max.hosts* and *Min.hosts* specify the range of changes in the number of active hosts during the simulation process[4]. The frequency of appearing and disappearing resources is defined by *Add host* and *Delete host*, according to the constant distributions for the static case, and normal distributions in the dynamic case. The initial number of tasks is denoted by *Init. tasks*parameter, which is constant in the static case. New tasks in the dynamic scheduling can be submitted to the system with the frequency defined by *Interarrival* parameter until the *Total tasks* value is reached. The *Scheduler strategy* parameter defines the type of scheduler, the termination condition for the scheduler and the optimization mode for the scheduling objectives. The setting $GA\_Scheduler(25,s)$ means that the simulator runs the GA-based scheduler for 25 seconds in simultaneous optimization mode[5].

2.3.3 Heuristic Schedulers Integrated with the Simulator

The *Sim-G-Batch* simulator allows and facilitates an integration of a mixture of heuristic scheduling algorithms. The simulator architecture enables to design the schedulers as the external dynamic-link libraries (dll files) and store them separated from the simulator main body. The schedulers are plugged in the simulator by using and *Adapter* pattern as it is presented in Fig. 2.2.

The heuristic scheduling methods are usually classified into three main groups, namely (1) calculus-based (greedy algorithms and ad-hoc methods); (2) stochastic (guided and non-guided methods); and (3) enumerative methods (dynamic programming and branch-and-bound algorithm). The heuristic schedulers integrated with *Sim-G-Batch* simulator are divided into three classes, namely *ad hoc*, *local search-based*, *population-based* heuristics. A simple taxonomy of those schedulers is presented in Fig. 2.3.

Ad-Hoc Methods

These methods are usually used for single-objective optimization. They are characterized by the low computational cost. Those methods are also very useful in

[4] In the case of dynamic scheduling, they are different from the initial number of hosts.

[5] The parameter h is used for hierarchical optimization mode , e.g. $GA\_Scheduler(25,h)$.

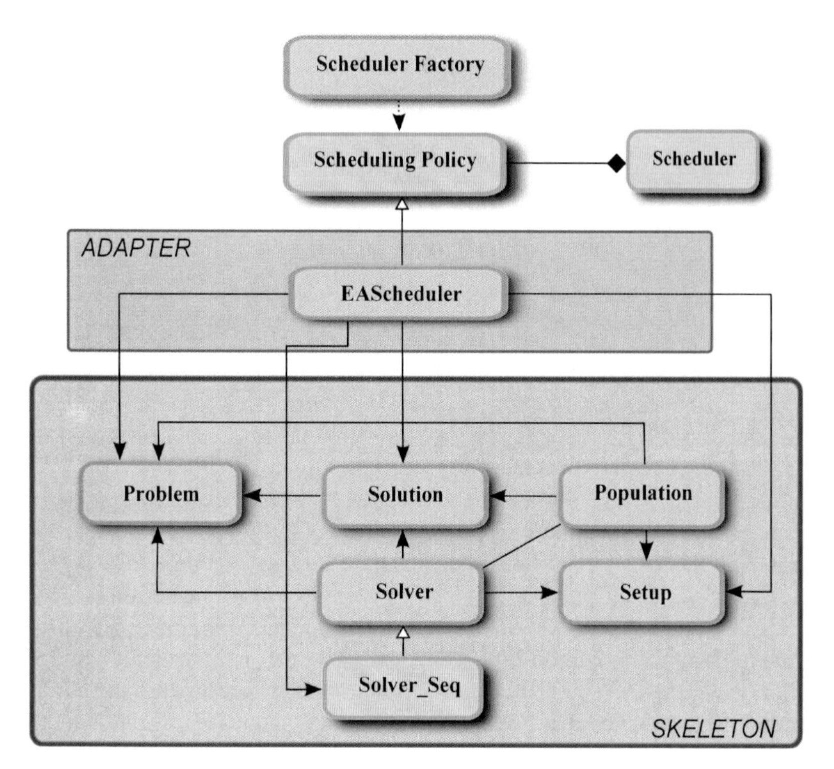

Fig. 2.2 The simulator adapter pattern used for different evolutionary based grid schedulers

generating the initial solutions for population-based schedulers. The Ad-hoc heuristics could be categorized as immediate mode heuristics and batch mode heuristics.

The *Immediate Mode Heuristics* group includes the following schedulers:

- *Opportunistic Load Balancing (OLB)*, sends a task to the first idle machine without taking into account the machines execution time;
- *Minimum Completion Time (MCT)*, assigns tasks to machines yielding the earliest completion times;
- *Minimum Execution Time (MET)*, assigns tasks to the machine having the smallest execution time for this task.

The *Batch Mode Heuristics* group contains the following methods:

- *Min-Min*: In this method for each task the machine yielding the earliest completion time is computed, then the task with the shortest completion time is selected and mapped to the corresponding machine.
- *Max-Min*: This method differs from the Min-Min as far as the final selection of the task with the latest completion time.
- *Sufferage*: The main idea of this method is to assign a given machine a task, which would "suffer" more if it were assigned to any other machine.

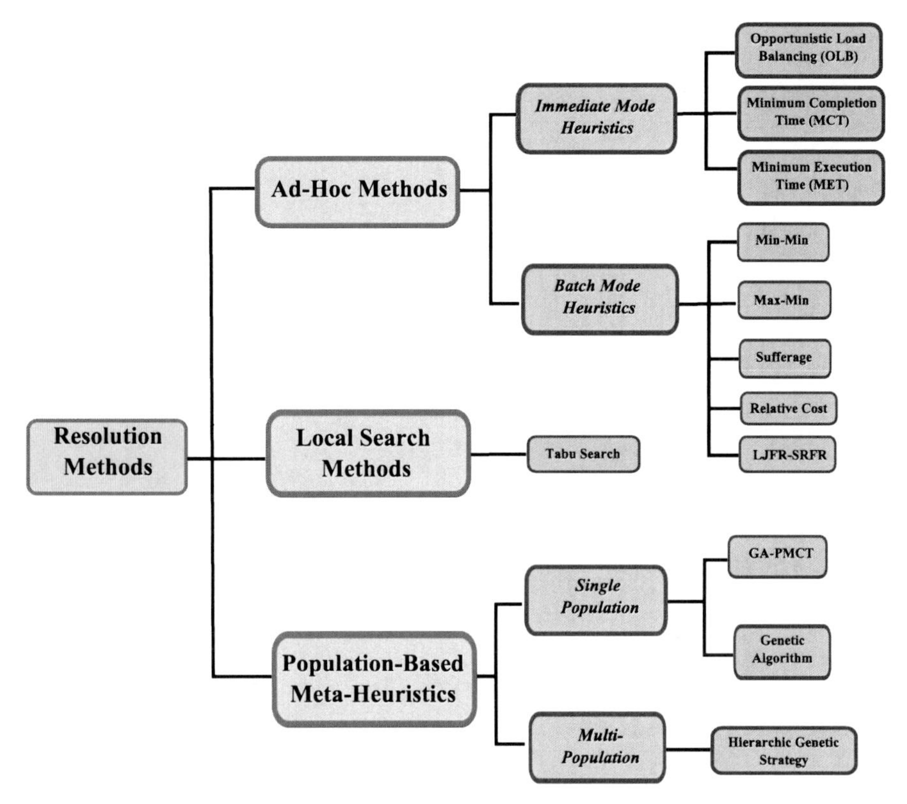

Fig. 2.3 Taxonomy of heuristic schedulers integrated with *Sim-G-Batch*

- *Relative Cost*: This methods allocates the grid resources according to the load balancing of machines and the execution times of tasks on machines.
- *Longest Job to Fastest Resource - Shortest Job to Fastest Resource (LJFR-SRFR)*: This method tries to simultaneously minimize both makespan and flowtime values: LJFR minimizes makespan and SJFR minimizes flowtime.

Local Search Methods

Methods from this category explore the optimization domain by constructing a sequence ('path') of partial solutions in optimization space. The most effective local-based grid scheduler is Tabu Search (TS) [162]. TS can be easily hybridized with more sophisticated schedulers (like GAs) to improve their efficiency.

Population-Based Heuristics

These methods use a population of individuals for the exploration of the solution space. The individuals in the populations represent the partial solutions of the considered problem. The major group in this class is a mixture of the single population

Genetic Algorithms (GAs), proposed by many authors as the effective grid schedulers [160, 161, 120]. Recently, a multi-population hierarchical GA-based scheduler has been defined in [90], [86]. In this method a set of dependent genetic processes is executed simultaneously. Each process creates a branch within the whole strategies tree structure, by using the GA-based scheduler with different settings. The search accuracy in a given branch (expressed as the branch degree parameter) depends on the mutation probability set for the scheduler activated in this branch (the higher mutation problem–the lower accuracy). A generic model of the hierarchical genetic scheduler and its several implementations are presented in Chapters 3-8 in this book.

Part II
Multi-Level Genetic-Based Hierarchical Grid Schedulers

Chapter 3
A Multi-Level Genetic Scheduling in Dynamic Grid Reinforced by the Population Hierarchy: Basic Model

Abstract. Exploring the search space in grid scheduling is a very complex problem, mainly because of the sheer size of the space and the dynamics of the system. This chapter presents a general concept of multilevel genetic metaheuristic scheduling, that is easily implemented and adapted to the various scheduling scenarios in the dynamic grid environment.

3.1 Introduction

Formal analyses of optimization landscapes for many classical combinatorial problems, such as *Traveling Salesman Problem* , *Graph Bi-Partitioning* , *Flowshop* scheduling, etc. [138], are potentially the outline for the development of mathematical models for the wide range of *NK* family optimization landscapes. This chapter allows for the defining and estimating of the distributions of the solutions. The features of the optimization landscape depend on the resolution methods used for problem solving. In contrast, the detailed characteristic of such landscapes allows individuals to tune the optimizer configuration to adapt to search mechanisms, particularly to the instance of the problem. In grid scheduling, this modelling is much more complicated, mainly because of different local scheduling policies and the system dynamics [100].

As a result of the wide assortment of constraints and different optimization criteria in the grid scheduling, heuristic and metaheuristic methods are the most feasible solutions for the grids scheduling problems. Metaheuristic schedulers Can easily explore the robustness of the search space. Another benefit of the scheduler is its ability for tackling various scheduling attributes, such as immediate and batch scheduling, multi-objectivity, decentralized and hierarchical grid architecture, etc [19].

Most of the currently available metaheuristic algorithms attempt to find an optimal solution with respect to a specific fitness measure. In the case of Genetic Algorithms (GAs) a great deal of effort has gone into designing efficient representation schemes and genetic operators so as to produce capable and effective solutions. The

J. Kołodziej: Evolutionary Hierarchical Multi-Criteria Metaheuristics, SCI 419, pp. 33–43.
springerlink.com

major challenges when using GAs to solve dynamic optimization problems are the ability to maintain diversity (or generate diversity) in the population and the ability to create robust solutions that are able to track the global solutions of the scheduling problem.

Table 3.1 presents selected examples of major projects in GA-based Grid scheduling. The presented algorithms belong to a wide class of *population-based* schedulers, and are divided into two additional categories, namely *single-population* and *multi-population* metaheuristics.

Table 3.1 Population-based metaheuristics in grid scheduling

Metaheuristic class	Class characteristic	Scheduler type	Methods
Population-based	- explore of the search space by the populations of individuals - require a large running time - effective in finding near-optimal solutions	Single-population	*Genetic Algorithms (GAs)* *Memetic Algorithms (MAs)* *Particle Swarm Optimization (PSO)* *Ant Colony Optimization (ACO)*
		Multi-population	*Island Genetic Algorithm* *Grid-Enabled Hierarchical Parallel Genetic Algorithm (GE-HPGA)*

Single-population GAs grid schedulers are presented in several works on grid computing. Zomaya and Teh [168] used GAs for dynamic load balancing. Braunt et al. [20] compare the efficiency of a simple GA-based scheduler based on methods from the set of ten static meta-task mapping heuristics from literature, including Min-Min, Min-Max, Minimum Completion Time (MTC) algorithms [7]. The authors provided their empirical study for the static benchmark for independent job scheduling in distributed heterogenous computing environment. The instances in this benchmark have been defined using the Expected Time to Compute (ETC) matrix model [5]. The same type of scheduling problem is considered by Xhafa et al. [161]. The authors examine in their study several combinations of GAs operators in order to identify the configuration of most effective and efficient operators and parameters for the problem. Following this the efficiency of the GA-based scheduler with the most effective and efficient combination of operators is compared to the effectiveness of the GAs approach presented in [20]. This model was extended by plugging the GA scheduler into a grid simulator [163] in order to perform the experiments in a dynamic Grid environment. The results of the evaluation of the GA scheduler were reported in [160]. Other GA approaches to different problems can be found and addressed by Martino and Mililotti [37], Page and Naughton [116], Gao et al. [19]. Another class of population-based methods is *Memetic Algorithms (MAs)*. These algorithms combine the genetic algorithms (or evolutionary strategies) with the local search methods. Therefore, MAs could be considered as hybrid evolutionary algorithms. What is more so is that MAs occurred as an attempt to combine concepts and strategies from different meta-heuristics. Xhafa in [156] applied unstructured MAs for independent scheduling under the ETC model. Other examples can be found in [31].

The implementation of the Ant Colony Optimization (ACO) algorithm to the grid scheduling problem modelled by the ETC matrix has been reported by Chang et al. in [30]. Lorpunmanee et al. in [99] applied ACO to the dynamic scheduling in grids. Other single-population approaches to the grid scheduling include the Particle Swarm Optimization [96], economic-based approaches [24] and Artificial Immune Systems [143]. A detailed survey of genetic-based meta-heuristics in grid scheduling is presented in [91].

In contrast to the single-population meta-heuristics, multi-population class of genetic-based grid schedulers is very small. The major projects include Grid-Enabled Hierarchical Parallel Genetic Algorithm (GE-HPGA) proposed by Lim et al. [95]. In this model the parallel single-population GAs, which generates the partial solutions of the scheduling problem, are managed by the centralized scheduler in a Master-Slave configuration. Rubio-Solar et al. [127] propose a *Island Genetic Algorithm* for solving the placement and routing problems in grids.

One of the most important and beneficial features of meta-heuristics in grid scheduling is that they can be easily hybridized with other approaches. It makes the receptive Grid schedulers adaptive to the various Grid types and specific types of applications. Abraham et al. [2] present a model for the hybridization of GA, Simulated Annealing (SA), and Tabu Search (TS) heuristics. Each GA-based hybrid, namely GA+SA and GA+TS, improves the efficiency of the genetic scheduler. Ritchie and Levine [124] combine an ACO with a TS algorithm for Grid scheduling.

3.2 Hierarchic Genetic Strategy Based Scheduler (*HGS-Sched*)

The exploration of the search space in combinatorial optimization remains a challenging problem, because of the sheer size of this space. In grid scheduling the task is even more difficult due to the high parameterization of the optimization problem as well as system dynamics. In addition, while the list of all possible schedules is determined by the permutations of tasks' or machines' labels (see Sec. 2.2.1), it is important to note that the lengths of these permutation strings may vary as the number of tasks and machines can change over time.

This section presents the general concept of the Hierarchic Genetic Scheduler (*HGS-Sched*). *HGS-Sched* framework is based on searching the grid environment through the execution of many dependent evolutionary processes.

3.2.1 HGS-Sched *Essentials*

The main concept of *HGS-Sched*'s architecture is based on the general model of Hierarchic Genetic Strategy (HGS) developed by Kołodziej et al. [80] [79] for global optimization in continuous domains. In the simplest implementation of HGS, the strategy is modelled as a decision tree and the individuals in each population are encoded using binary strings of different lengths. In such cases a single-population GA is applied as a main genetic mechanism at all levels of the strategy [129]. In another implementation of HGS [154], the main mechanism in all branches of the tree has been defined by using various evolutionary strategies with populations of individuals encoded by the vectors with the floating-point coordinates.

HGS algorithms span few classes of the parallel GAs taxonomies (see [142], [4]). For example, according to the classification presented in [112], HGS algorithms can be categorized as a 'Dynamic Deme'-like method and a method from the class of meta-heuristics with the 'adaptive accuracy of search'.In fact, the HGS framework may be easily adapted for designing of many single-population GAs. However, as a result it is very difficult to find proper classification rules for this strategy.

HGS strategy has been successfully applied as an efficient method for Permutation Flowshop Scheduling [83] , as well as for solving some practical engineering problems [81], [84]. Both basic HGS models can be used for solving combinatorial global optimization problems. However, the strategy in this case may generate many 'infeasible' solutions to the problem and may need some additional specialized repair algorithms.

In *HGS-Sched* the search process is initialized by activating a scheduler with low search accuracy. This scheduler is the main module of the entire strategy and is responsible for the 'management' of the general structure of the tree[1] and exploration of new and unrecognized regions in the optimization domain. As a result of its low accuracy, the main scheduler is not very effective in detecting the global solutions to the problem. However, it may generate in a short amount of time a large number partial and potential solutions, which can be validated by the activation of more accurate processes The activation of these processes does not dramatically increase the complexity of the whole hierarchical scheduler for the following three reasons:

- In contrast to the hybrid strategies, where the components are usually composed of various meta-heuristics and local search methods, the same general framework is applied for modelling the algorithms working at all levels of the tree.
- The management of the tree structure is steered by specialized operations responsible for the deactivation of the flawed processes.
- Finally, the synchronization of the search is provided 'horizontally' at each level of the tree, so there is no need to refer to the tree's parental nodes. This feature enables an easy adaptation of the strategy to the current state of the grid system.

For all of the above reasons *HGS-Sched* significantly differs from all existing hierarchical, hybrid, and branching schedulers applied for solving the grid scheduling problems and classical job-shop problems (see e.g. [21]).

Figure 3.1 depicts a simple graphical representation of 3-level structure of *HGS-Sched*.

[1] The scheduler with the lowest accuracy of search is called *the core* of the HGS tree structure.

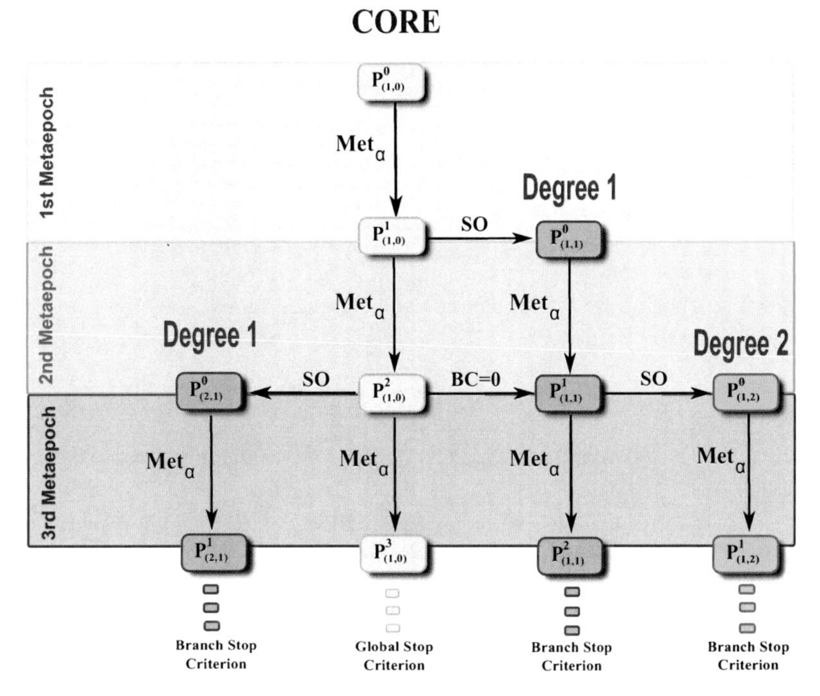

Fig. 3.1 3 levels of *HGS-Sched* tree structure

The configuration of the *HGS-Sched* strategy is defined by the following key parameters:

- t – degree of the branch;
- $\widetilde{Max}$– maximal degree of the branches;
- $1 \geq pop_1 \geq pop_2 \geq \ldots \geq pop_{\widetilde{Max}}$ – cardinalities ('sizes') of populations in the branches;
- $\mu_1 \geq \mu_2 \geq \ldots \geq \mu_{\widetilde{Max}}$ – mutation parameters (rates) in the branches;

Each branch of the tree is created by an active GA designed for solving scheduling problems. The accuracy of search in *HGS-Sched* branches is defined by the *degree* parameter with the lowest value, e.g. 0, set for the core of the system[2].

Populations of the schedules (individuals) are the main structures in the *HGS-Sched* branches. A population in a branch of degree t is denoted by $P^{\bar{e}}_{(r,t)}$, where:

- $\bar{e} \in \mathbb{N}$ defines the global metaepoch counter,
- r is the number of branches of the same degree in the tree.

The hierarchical structure of the scheduler is modified periodically after the execution of α-generation evolutionary processes in each active branch. Such a process is called a *α-periodic metaepoch* Met_α, $(\alpha \in \mathbb{N})$ and define it in the following way:

[2] The *HGS-Sched* framework may be used for the implementation of the single-population genetic algorithms and evolutionary strategies. In such a case the value of t parameter is 0.

Definition 3.1. A α-periodic metaepoch $Met_\alpha, \alpha \in \mathbb{N}$ is a discrete evolution process which starts by selection of the best adapted individual from the given population and terminating after α generations.

$$Met_\alpha\left(P^{\bar{e}}_{(r,t)}\right) = \left(P^{\bar{e}+l}_{(r,t)}, \widehat{S}\right); \tag{3.1}$$

where $\widehat{S}$ is the best adapted individual in the metaepoch, $P^{\bar{e}}_{(r,t)}, (t \in \{0,\ldots,\widetilde{Max}\}, \widetilde{Max} \in \mathbb{N})$.

New branches of the higher degree can be created by using a *Sprouting Operation (SO)* defined as follows:

$$SO(P^{\bar{e}}_{(r,t)}) = (P^{\bar{e}}_{(r,t)}, P^{0}_{(r',t+1)}), \tag{3.2}$$

where $P^{\bar{e}}_{(r,t)}$ is a parental branch, and $P^{0}_{(r',t+1)}$ denotes the initial population for a new branch of degree $t+1$.

Individuals of this population are selected from an S_t-neighborhood ($1 \leq S_t \leq n$) of the best adapted individual $\widehat{S}$ in the parental population $P^{\bar{e}}_{(r,t)}$. This neighbourhood is defined as the set of modifications of the schedule $\widehat{S}$ created by all possible permutations or reassignments of tasks in $(n - S_t)$-length suffix of $\widehat{S}$. The S_t-length prefix of a given schedule S is generated by using the following operator:

$$A_{(S_t)}(S) = \tilde{S}, |\tilde{S}| = S_t, S_t \leq n \tag{3.3}$$

where $|\tilde{S}|$ denotes the length of the suffix in the permutation sequence which encodes the schedule S. The values of S_t parameters may be different in branches of different degrees. In this approach they are calculated in the following way:

$$S_t = (suf)^t \cdot n, \tag{3.4}$$

where $suf \in [0,1]$ is a global strategy parameter called a *neighborhood parameter* and t is the branch degree.

The search process in all sprouted branches must be more detailed than in the parental ones. In *HGS-Sched* the accuracy of the search is determined by the value of the mutation probability. This parameter should be more refined in the sprouted branches than in the parental branch. The process extending the tree in *HGS-Sched* is different than the sprouting mechanism in the binary implementation of HGS [79]. In binary implementation the chromosomes of the individuals in the sprouted branches are interpreted as nodes of the encoding mesh, which is denser that in the case of the encoding mesh defined for the parental branch. The density of the encoding structure defines the accuracy of the search in the HGS branches. This encoding method is called *Nested Coding* [129].

The sprouting operation is conditionally activated depending on the results of the application of a *Branch Comparison* (*BC*) binary operator to the parental branch and its all directly sprouted branches. This operator is used for the detection of 'similarity' of the resulting populations in each parental-sprouted pair of branches. Formally the $BC : POP \to \{0,1\}$ operator is defined by the following formula:

$$BC(P,\hat{P},S_t) = \begin{cases} 1, \; \exists S \in P, \exists \hat{S} \in \hat{P} : A_{S_t}(S) = A_{S_t}(\hat{S}) \\ 0, \; \text{otherwise,} \end{cases} \tag{3.5}$$

where $POP = \{(P,\hat{P},S_t)\}$ and $P,\hat{P}$- are populations within branches of degrees t and $t+1$ respectively. This operator is activated after execution of at least two metaepochs in the core.

The outcome of the BC operator is 1 if the parental branch and its 'descendant' (sprouted) branch operates in a similar region within the optimization landscape. In such a case, another meta-epoch is executed in the parental branch without creating a new process. This technique is crucial for the effective management of the algorithm structure and preventing the activation of many similar processes in the same local region, which could significantly increase the complexity of the strategy as a whole.

The execution of the BC operator can be time consuming, as such as in the implementations of the prototypes of *HGS-Sched* in static networks (see [90]). In those early versions of the hierarchical scheduler all populations at a given tree level were scanned and individuals with the same prefixes were found. In the current version of *HGS-Sched* the *hash technique* is implemented in order to reduce the execution time of the procedure BC. The hash table is defined with the *task-resource allocation* key denoted by K. The value of this key is calculated as the sum of the absolute values of the subtraction of each position and its precedent in the S_t-length suffix in direct representation of the schedule vector (reading the suffix in a circular way). The hash function f_{hash} is defined as follows:

$$f_{hash}(K) = \begin{cases} 0, & K < K_{min} \\ \left\lfloor N \cdot \left(\frac{K - K_{min}}{K_{max} - K_{min}} \right) \right\rfloor & K_{min} \leq K < K_{max} \\ pop - 1, & K \geq K_{max} \end{cases} \tag{3.6}$$

where K_{min} and K_{max} correspond respectively to the smallest and the largest value of K in the population, and pop is the population size.

In the case of the conditional sprouting of new branches of the degree $t+1$ from the parental branch of the degree t the keys are calculated for the best individual in the parental branch and individuals in all populations in all active branches of the degree $t+1$. If there is any individual in the higher degree branches for which the key matches the key of the best adapted individual in the parental branch, then the value of BC is 1 and no branch of the degree $t+1$ is sprouted.

In the case of the comparison of the branches of the same degree t, all branches in which there exists individuals with identical keys must be reduced and a single joint branch created (the value of BC is 1). The individuals in this branch are selected from the 'youngest' (in the sense of the population evolution) populations in all reduced branches.

3.3 Genetic Mechanism in *HGS-Sched* Branches

Several single-population genetic algorithms may be implemented as the main genetic mechanism in the *HGS-Sched* branches [11]. However, it is recommended to use simple mechanisms in order to keep the complexity of the whole strategy at a lower level. Alg. 1 defines a generic framework for a genetic engine implemented in all *HGS-Sched* branches for the experiments described in this book. This framework is based on the general model of the conventional single-population GA [107].

Algorithm 1. A template of the genetic engine for *HGS-Sched*

1: Generate the initial population P^0 of size μ; $e = 0$
2: Evaluate P^0;
3: **while** not termination-condition **do**
4: Select the parental pool T^e of size λ; $T^t := Select(P^e)$;
5: Perform crossover procedure on pars of individuals in T^e with probability p_c; $P^e_c := Cross(T^e)$;
6: Perform mutation procedure on individuals in P^e_c with probability p_m; $P^e_m := Mutate(P^e_c)$;
7: Evaluate P^e_m ;
8: Create a new population P^{e+1} of size μ from individuals in P^e and P^e_m ; $P^{e+1} := Replace(P^e; P^e_m)$
9: $e := e+1$;
10: **end while**
11: **return** Best found individual as solution;

The parameter e in Alg. 1 is the counter of the generations in GA (e counts the number of loops in GA). The *direct representation* method defined in Sec. 2.2.1 (see Chapter 2) is used for encoding schedules in the base populations P^e and P^{e+1}, and *permutation representation* – in P^e_c and P^t_m populations.

The initial population in Alg. 1 is generated by using the *Minimum Completion Time + Longest Job to Fastest Resource - Shortest Job to Fastest Resource MTC + LJFR-SJFR* method, in which all but two individuals are generated randomly. Those two individuals are created by using the *Longest Job to Fastest Resource - Shortest Job to Fastest Resource (LJFR-SJFR)* and *Minimum Completion Time (MCT)* heuristics [161]. In LJFR-SJFR method initially the number of m tasks with the highest workload are assigned to the available m machines sorted in ascending order by the computing capacity criterion. Then the remaining unassigned tasks are allocated to the fastest available machines. In the MCT heuristics, a given task is assigned to the machine yielding the earliest completion time. The detailed definition of those procedures may be found in [29].

Alg. 1 was adapted to the CG scheduling problem through an implementation of specialized encoding methods and genetic operators. The operators from the following set were used in experiments presented in this book:

- **Selection operators:** *Linear Ranking*;
- **Crossover operators:** *Partially Mapped Crossover (PMX)* and *Cycle Crossover (CX)*;
- **Mutation operators:** *Move*, *Swap* and *Rebalancing*;
- **Replacement operators:** *Steady State*, *Elitist Generational*, *Struggle*.

All the above mentioned operators are commonly used in the genetic meta-heuristics dedicated to solving combinatorial optimization problems [36]. The detailed definition and examples may be found in [11].

In *Linear Ranking* method a selection probability for each individual in a population is proportional to the rank of the individual. The rank of the worst individual is defined as zero, while the best rank is defined as $pop\_size - 1$, where pop_size is the size of the population.

The following crossover and mutation operators are implemented for permutation-based representation of the schedules. In the examples we show the results of the crossover just on the vectors u (see Chapter 2, Sec. 2.2.1).

In *Partially Matched Crossover (PMX) [55]* a segment of one parent-chromosome is mapped to a segment of the other parent-chromosome (corresponding positions) and the remaining genes are exchanged according to the mapping 'relationship' of tasks and machines specified by the concrete scheduling rules. An example of the result of using the PMX operator for two schedules of the length 10 is defined as follows:

$$\begin{array}{ll} \textbf{Parents} & (10\ 9\ 6\ |5\ 3\ 7\ 8|\ 1\ 4\ 2) \\ & (10\ 5\ 3\ |7\ 4\ 1\ 8|\ 2\ 6\ 9) \\ \textbf{Offsprings} & (10\ 9\ 6\ |7\ 4\ 1\ 8|\ 5\ 3\ 2) \\ & (10\ 1\ 4\ |5\ 3\ 7\ 8|\ 2\ 6\ 9) \end{array} \tag{3.7}$$

Tn *Cycle Crossover (CX)* [113], first, a cycle of alleles is identified. The crossover operator leaves the cycles unchanged, while the remaining segments in the parental strings are exchanged. An example of the result of using the CX operator for two schedules of the length 10 from the previous example is defined as follows:

$$\begin{array}{ll} \textbf{Parents} & (10\ 9\ 6\ 5\ 3\ 7\ 8\ 1\ 4\ 2) \\ & (10\ 5\ 3\ 7\ 4\ 1\ 8\ 2\ 6\ 9) \\ \textbf{Cycles} & (-\ 9\ -\ 5\ -\ 7\ -\ 1\ -\ 2) \\ & (-\ 5\ -\ 7\ -\ 1\ -2\ -\ 9) \\ \textbf{Offsprings} & (\ 10\ 9\ 3\ 5\ 4\ 7\ 8\ 1\ 6\ 2) \\ & (\ 10\ 5\ 6\ 7\ 3\ 1\ 8\ 2\ 4\ 9) \end{array} \tag{3.8}$$

In *Move* mutation a task is moved from one machine to another one. Although the task can be appropriately chosen, this mutation strategy tends to unbalance the number of jobs per machine. It is realized by the modification of two coordinates in the vector $\tilde{u}$ of the schedule code (see Chapter 2, Sec. 2.2.1).

The main idea of the *Rebalancing* method is to first improve the solution (by rebalancing the machine loads) and then mutate it. A rebalancing procedure is executed as follows. First, the most overloaded machine is selected. Two tasks j and $\hat{j}$

are identified in the following way: j is assigned to another machine i', $\hat{j}$ is assigned to i and $ETC[j][i'] \leq ETC[\hat{j}][i]$. Then the assignments are interchanged for tasks j and $\hat{j}$.

In *Swap* mutation the indexes of two selected tasks in the schedule representation are swapped.

A base population for a new GA loop in Alg. 1 may be defined by using the *Elitist generational* replacement method, where a new population contains two best solutions from the old base population and the rest are the newly generated offsprings.

In the *Steady State* replacement method, the set of the best offsprings (the number of elements in this set is fixed) replaces the worst solutions in the old base population. The main drawback of this methods is that it can lead to the premature convergence of the algorithms in some local solutions.

The *Struggle* replacement mechanism can be an effective tool for avoiding too fast of a scheduler's convergence to the local optima. In such method, new generations of individuals are created by replacing a part of the population by the individuals most similar – if this replacement minimizes the fitness value. The definition of the struggle replacement procedure requires a specification of the appropriate *similarity measure*, which indicates the degree of the similarity among two GA's chromosomes. We use in this work the *Mahalanobis distance* [101] for measuring the distances between schedules according to the following formula:

$$sim_e(S^1;S^2) = \sqrt{\sum_{j=1}^{n} \frac{(S^1[j] - S^2[j])}{\sigma_P^2}} \tag{3.9}$$

where σ_P is the standard deviation of the $S^1[j]$ over the population P.

The struggle strategy has shown to be very effective in solving several large-scale multi-objective problems (see e.g., [14], [59]). However, the computational cost can be very high, because of the need to calculate distances among all off springs in resulting population and the individuals in the base population for the current GA loop. To reduce the execution time of the struggle procedure we use a *hash technique*, in which the hash table with the *task-resource allocation* key is created. The value of this key, denoted by K, is calculated as the sum of the absolute values of the subtraction of each position and its precedent in the direct representation of the schedule vector (reading the schedule vector in a circular way). The hash function for the struggle replacement method is defined by using Eq. (3.6). The struggle mechanism allows a tuning of the Grid scheduler to the particular scheduling problem and scenario.

3.4 Summary

The general idea of adaptive exploration of the search space in global optimization is a basic principle of evolutionary computation [54]. The accuracy of the search process is usually defined by the crossover and mutation parameters [11]. These parameters may be dynamically adapted to the current state of the algorithms or the optimization landscape. However, the high accuracy in the exploration of the optimization landscape is achieved by hybrid techniques with the main genetic mechanism and other optimization techniques (local search method or another meta-heuristic) [42], or Island genetic models [153]. In both cases some prior knowledge of the optimization landscape is needed. This knowledge is not necessary for *HGS-Sched* model presented in this chapter. In conventional scheduling problems, such as static or dynamic job-shop scheduling, the optimization landscape for a single-objective scheduling (with a makespan as a scheduling criterion) is characterized by Reeves in [123] . Reeves defined a landscape structure for the small-size problems (5 machines and 1—20 tasks) as the set of separated clusters of local solutions. The sizes of the clusters in this model are rather small, and the distances between the different clusters – significant. In grid scheduling such characteristic may be very complex. In fact, there are no theoretical models that can tackle the high complexity and dynamics of the grid systems and all relations among the grid users. Additionally, the Markov models of GAs [148] and HGS algorithms [79] cannot be easily adapted to the scheduling in grids. In such models the knowledge of all system states is necessary, which makes the models ineffective in the large-scale dynamic grid environment. The following chapters in this book will present a comprehensive empirical study of various implementations of HS-Sched in different scheduling scenarios.

Chapter 4
Hierarchic vs. Single–Population and Hybrid Metaheuristic Grid Schedulers: A Comparative Empirical Study

Abstract. This chapter presents the results of comprehensive empirical evaluation of hierarchical, hybrid, single- and multi-population genetic metaheuristics in static and dynamic versions of the scheduling problem in grid. All metaheuristics have been integrated with the *Sim-G-Batch* grid simulator. The results of the analysis show the high effectiveness of *HGS-Sched* in exploration of the bi-objective dynamic optimization landscapes in highly-parametrized grids.

4.1 Introduction

The formal analysis of the features of genetic-based meta-heuristics remains a research challenge from the early studies on Evolutionary Computation (EC). While all genetic and evolutionary-like methodologies has grown far beyond the original Genetic Algorithm (GA) concept defined by John Holland [63], the studies on the theoretical models of the GA-based techniques still cannot extend significantly the basic models of binary coded GAs [148], [149], [153] and (1+1)-like evolutionary strategies [128].

The first reason of the stagnation of this kind of research may be the complexity of the optimization process driven by the metaheuristics. The metaheuristics try to generate high quality solutions of the problem by making a series of improvements during their iterative process. Whether they start with randomly generated low quality solutions or they use smart initialization methods to take advantage of the problem-specific knowledge, they aim to improve solution quality during the search process. At any iteration a metaheuristic method must make some decisions about loosing or keeping the partial solutions for further processing. This decisions are often responsible for maintaining a balance between exploration and exploitation of the search space. Too radical decisions may reduce the exploratory capabilities of the algorithm and often results in a premature convergence. In the case of genetic and evolutionary algorithms this problem is even more complicated because of many selection procedures that must be executed in just the single iteration of the algorithm for selection of the parental mid-population and the base populations. These

J. Kołodziej: Evolutionary Hierarchical Multi-Criteria Metaheuristics, SCI 419, pp. 45–77.
springerlink.com

operations must be synchronized with variation operators and constraint-handling methods. Unfortunately, any formal GAs model, and in fact also algorithm's framework, to be successfully applied, must be tailored to a specific domain in order to be useful and provide decent results in specified time. Therefore, the development of a generic formal model for a wide class of the metaheuristics for the combinatorial optimization is in fact almost impossible. Moreover, the simple models usually fail in the case of the dynamic changes of problem's and system's settings. In this case the best method of the verification of the features of meta-heuristics is their empirical evaluation.

The empirical analysis of metaheuristics in the large-scale dynamic environment needs a specification of a large set of parameters for the environment. In order to ensure the achievement of valuable and statistically significant results, the methodologies must be tested on a 'representative', which also means 'large', set of problem instances.

For the static optimization problems, the experimental evaluation and the tuning of parameters is done through benchmarks of (static) instances. The main aim of the analysis in this case is to run the metaheuristic a sufficient number of times on the same instance and using a fixed setting of parameters in order to compare the results with the best results achieved by other well-known meta-heuristics. For some classical combinatorial problems those results are stored as reference values in OR-Library [114][1]. In order to avoid biased results, the fine-tuning of parameters is provided.

In the dynamic case, the set of strategic parameters of the system is larger than in the static scenario. The dynamic systems are in fact the decision-support systems that require continuous flow of data, predictive components, almost immediate recommendations for recovering from violent changes, etc. The optimization problems defined for such systems usually deal with many variables, nonlinear relationships, huge variety of constraints, business rules, many (usually conflicting) objectives and all of these are set in a dynamic and noisy environment. Any empirical analysis in such complex scenarios may be very expensive for all system users, and in some cases, simply impossible. Therefore simulation seems to be the key methodology also for the empirical comparative analysis of the metaheuristics used for solving the complex optimization problems in highly-parametrized dynamic environments.

The goal of the empirical analysis presented in this chapter is the comparison of the effectiveness of single- and multi-population metaheuristics, namely *HGS-Sched* defined in Chapter 3, numerous variants of single-population GA-based schedulers, and hybridized *GA+ TabuSearch* solution, in the independent batch scheduling with minimizing the makespan and flowtime as the scheduling objective functions. These functions are optimized in the hierarchical mode with the makespan as the most important criterion. All tested metaheuristics are integrated with the *Sched-G-Batch* grid simulator (see Chapter 2, Sec. 2.3), which allows to evaluate the schedulers in the static and dynamic scheduling scenarios on a big set of benchmarks.

[1] OR-Library was originally specified by J.E.Beasley in [15].

All experiments are scheduled as follows. In the first part of the analysis the HGS-Sched method is tested along with the conventional and best so far single-population GA schedulers in the small size static scheduling by using the benchmark generated by the ETC Matrix model. The main aim of those experiments is to verify and compare the possible impact of heterogeneity of the tasks and grid resources on the efficiency of the schedulers in the balancing of the loads of the grid machines, and as the result, the minimization of the makespan and flowtime criteria. The results of these experiments are presented in Sec. 4.3. In the second part of the analysis eights variants of single-population GAs, HGS-Sched and $GA+TS$ hybrid are evaluated in the large-scale static and dynamic grid scenarios. The results of these experiments are presented in Sec. 4.4.

4.2 The Settings of the Grid Simulator and Scheduler Performance Measures

The instances of the scheduling problems for the empirical study presented in this chapter are generated by the *Sim-G-Batch* simulator defined in Chapter 2 (see Sec. 2.3). The basic set of the input data for the simulator includes:

- the workload vector of tasks,
- the computing capacity vector of machines,
- the vector of prior loads of machines, and
- the ETC matrix of estimated execution times of tasks on machines.

The *Sim-G-Batch* simulator is highly parametrized to reflect the various realistic grid scenarios. The sample values of key input parameters used in the experiments for the simulator are presented in Table 4.1[2].

These parameters are interpreted as the global characteristics of the conventional grid systems with the low and high heterogeneities of the tasks and resources, and low and high system dynamics. The following four grid size scenarios are considered in the study: (a) 'Small' grid (32 hosts/512 tasks), (b) 'Medium' grid (64 hosts/1024 tasks), (c) 'Large' grid (128 hosts/2048 tasks), and (d) 'Very Large' grid (256 hosts/4096 tasks). The similar grid characteristics are used for an experimental analysis of the effectiveness of the heuristic schedulers in many research projects in grid computing [159], [29], [157], and in the recent works of the author of this book [82], [83], [86].

4.2.1 The Measures of the Schedulers Performance

The relative performances of all schedulers in experiments presented in this chapter were evaluated according to the following two criteria:

- *Makespan* – the primarily scheduling criterion defined in Eq. (2.15);

[2] The following notation $U[x,y]$, $N(a,b)$ and $E(c,d)$ is used for uniform, Gaussian and exponential probability distributions, respectively.

Table 4.1 Values of key parameters of the grid simulator in static and dynamic cases

	Small	Medium	Large	Very Large
		Static case		
Nb. of hosts	32	64	128	256
Resource cap. (in MHz CPU)		$N(5000, 875)$		
Total nb. of tasks	512	1024	2048	4096
Workload of tasks		$N(250000000, 43750000)$		
		Dynamic case		
Init. hosts	32	64	128	256
Max. hosts	37	70	135	264
Min. hosts	27	58	121	248
Resource cap. (in MHz CPU)		$N(5000, 875)$		
Add host	$N(625000, 93750)$	$N(562500, 84375)$	$N(500000, 75000)$	$N(437500, 65625)$
Delete host		$N(625000, 93750)$		
Init. tasks	384	768	1536	3072
Total tasks	512	1024	2048	4096
Inter arrival	$E(7812.5)$	$E(3906.25)$	$E(1953.125)$	$E(976.5625)$
Workload		$N(250000000, 43750000)$		

- $Mean\_Flowtime$ – mean flowtime calculated as follows:

$$Mean\_Flowtime = F \tag{4.1}$$

where F denotes the flowtime defined in Eq. (2.17).

4.3 The Evaluation of HGS-Sched on Static Benchmark for the Small-Size Grid

In this section *HGS-Sched* algorithm is evaluated in the static grid scenario for a small cluster of machines and small batch of tasks. The main aim of this study is to verify the impact of the distribution of tasks and resources in the system on the scheduler performance, and the effectiveness of the single- and hierarchical genetic meta-heuristics in the reduction of the system overheads. *HGS-Sched* algorithm is compared with the most representative single-population GA-based schedulers designed for the independent batch scheduling in grids.

4.3.1 The Benchmark Description

The benchmark for small static grid was generated by the *Static ETC Generator* module of the *Sim-G-Batch* simulator (see Sec. 2.3.1 in Chapter 2). The instances in this benchmark are classified into 12 types of ETC matrix, according to task heterogeneity, machine heterogeneity and consistency of computing. These instances are labeled by the following parameters

$$gamma\_xx\_yyzz.0 \tag{4.2}$$

where:

- *gamma* denotes the gamma distribution used in generating the ETC matrix;
- *xx* stands for the type of consistency of ETC matrix($\hat{c}$–consistent, $\hat{\bar{i}}$–inconsistent, and $\hat{s}$– semi-consistent);
- *yy* indicates the heterogeneity of tasks (*hi* – high heterogeneity, and *lo* – low heterogeneity);
- *zz* expresses the heterogeneity of the resources (*hi* – high, and *lo* –low).

All ETC matrices were generated by using the CVB method (see Chapter2, Sec. 2.2) with the following input parameters: $exec_{ave} = 10$, and $0.1 \leq tvar_{tasks}$, $mvar_{mach} \leq 0.35$. The grid cluster network is composed of 16 nodes (machines) and there are 512 tasks submitted for scheduling. The analysis starts by tuning the genetic engine for *HGS-Sched* algorithm. The result of this tuning process is a configuration of single-population GA algorithm, which is implemented as the genetic engine in *HGS-Sched*. *HGS-Sched* with such an engine is used in the comparison analysis of the performances of hierarchical meta-heuristic and two single-population GA-based schedulers defined in [20] and [161]. These single-population schedulers are promoted as the effective methods for solving small-scale static scheduling problems.

4.3.2 Tuning of GA Operators for HGS-Sched

The genetic operators used in the tuning process were selected from the set of operators (3.3) defined in Chapter 3, Sec. 3.3. Each experiment was repeated 30 times under the same configuration of operators and parameters and the average values of $Makespan$ and $Mean\_Flowtime$ were computed.

Table 4.2 shows the results achieved by the schedulers with two crossover operators, namely *Cycle Crossover (CX)* and *Partially Matched Crossover (PMX)*, and *Rebalancing* mutation. The *CX* crossover outperforms the rest of the operators in the minimization of both makespan and flowtime criteria.

Table 4.2 Comparison of crossover operators ([$\pm s.d.$]= standard deviation)

Operator	Average *Makespan*	Average *Mean_Flowtime*
PMX	8253658.817 [± 522982.428]	1083433775.989 [± 300917659.178]
CX	**7582290.438 [± 962174.801]**	**1032764993.452 [± 147902068.802]**

The results of the tuning the mutation operators are presented in Table 4.3. Three selected mutation methods –*Move*, *Swap* and *Rebalancing* – were combined with *CX* crossover operator. In this case *Rebalancing* mutation achieves the best results in the optimization of the scheduling objective functions.

Table 4.3 Comparison of mutation operators for makespan and flowtime values.

Operator	Average *Makespan*	Average *Mean_Flowtime*
Move	9201421.234 [± 719048.166]	1103421125.004 [± 109300376.601]
Swap	10753798.757 [± 12734428.834]	1197064425.190 [± 921729050.218]
Rebalancing	**7665531.021 [± 743666.66]**	**10416946991.943 [± 115207562.548]**

Three replacement operators were considered as the candidate methodologies of generation of the base populations for GA schedulers, namely *Steady State*, *Elitist Generational*, and *Struggle* operations. The results for both methods are presented in Table 4.4. The *Struggle* is the most effective in the optimization of the makespan and flowtime objectives although the differences in the results are minor.

Based on the results of this simple tuning process, the optimal configuration of genetic operators for the main engines of the GA-based grid schedulers is defined as follows:

- Selection – Linear Ranking;
- Crossover – CX;
- Mutation – Rebalancing;
- Replacement – Struggle.

This engine is used as the main genetic mechanism in all *HGS-Sched* branches in small-scale static scheduling.

Table 4.4 Comparison of performance of replacement operators.

Operator	Average *Makespan*	Average *Mean_Flowtime*
Steady State	7714176.552 [± 646120.259]	1054674832.098 [± 117471995.462]
Elitist Generational	7702158.131 [± 849364.304]	1040779095.420 [± 279901270.213]
Struggle	**7634563.365 [± 712499.473]**	**1022253853.537 [± 95836415.955]**

4.3.3 The Comparative Analysis of Single-Population and Hierarchical Genetic Schedulers in Static Scheduling in Small-Area Grid Cluster

The goal of the empirical analysis presented in this section is to compare the effectiveness of the hierarchical *HGS-Sched* algorithm with the results achieved by two representative single-population genetic meta-heuristics, namely the classical GA grid scheduler defined by Braunt et al. in [20], and Xhafa et al. [158]. The configuration of the Braunt's algorithm is defined by the set of the following genetic operators: random selection, . *OX* crossover, *Swap* mutation and *Elitist Generational* The configuration of the Xhafa's algorithm is identical with the optimal genetic engine of the *HGS-Sched* generated in the previous section.

The values of *HGS-Sched* control parameters are presented in Table 4.5. The *nb_of_metaepochs* denotes the maximal number of metaepochs executed in the core

Table 4.5 *HGS-Sched* global parameter values

Parameter	
degrees of branches (t)	0 and 1
period_of_metaepoch (α)	200
nb_of_metaepochs	10
neighborhood parameter (suf)	0.5
mut_prob(0)	0.4
mut_prob(1)	0.2
cross_prob	0.8

of the structure and it is defined as a global stopping criterion for the whole strategy. It is assumed that there are just two types of branches in *HGS-Sched* tree, namely core of the structure (one branch of degree 0 and sprouted branches of degree 1. The numbers of individuals in *HGS-Sched* populations were 50 individuals in the core and 18 – in the sprouted branches. The parameters $cross\_prob$, $mut\_prob(0)$ and $mut\_prob(1)$ are used for the notation of of the crossover and mutation probabilities.

In order to make a fair comparison analysis and reconstruct the experiments provided by Braunt and Xhafa a *work* parameter $Work$ was defined for all algorithms. This parameter expresses the amount of work of the particular scheduler. The $Work$ parameter is defined as follows:

$$Work = (mut\_prob + cross\_prob) \times pop\_size \times nb\_of\_generations \tag{4.3}$$

For *HGS-Sched* algorithm, the value of the $Work$ parameter is approximated by using the parameters in all possible branches, that is to say:

$$\begin{aligned} Work = \Big[&(mut\_prob(0) + cross\_prob) \times pop\_size(0) + \\ &+ \textstyle\sum_{t=1}^{\tilde{M}} \big((mut\_prob(t) + cross\_prob) \times \\ &\times pop\_size(t) \times nb\_br(t)\big)\Big] \times period\_of\_metaepoch \times nb\_of\_metaepochs \end{aligned} \tag{4.4}$$

where:

- $\tilde{M}$ is the maximal degree of all branches in *HGS-Sched*,
- $nb\_br(t)$ is the number of sprouted branches of degree t,
- $pop\_size(t)$ denotes the size of the population in the branch of degree t.

Based on Eq. (4.3) and Eq. (4.4), the amounts of work of three analyzed schedulers are specified as follows:

- for GA in [20]: $Work_{Braunt} = (0.4 + 0.6) * 200 * 1000 = 200000$,
- for GA in [158]: $Work_{Xhafa} = (0.4 + 0.8) * 68 * 2500 = 240000$, and
- for *HGS-Sched*: $Work_{HGS} = ((0.2 + 0.8) * 8 * 12 + (0.4 + 0.8) * 28) * 200 * 10 = 259200$.

It can be noted from the above calculation, that the values of work parameters are in the same range ensuring a proper comparison of the considered algorithms.

4.3.3.1 Results

Tables 4.6 and 4.7 present the average values of $Makespan$ and $Mean\_Flowtime$ achieved by the three considered GA-based schedulers[3].

Both *Makespan* and $Mean\_Flowtime$ metrics are expressed in arbitrary (but not concrete) time units.

[3] The flowtime criterion was not optimized in [20], but the configurations of operators and parameters were kept the same as in the case of makespan minimization.

Table 4.6 Comparison of average *Makespan* values for three hierarchical and single-population genetic schedulers ([±*s.d.*]: *s.d*–standard deviation)

Instance	Braunt's GA	Xhafa's GA	*HGS-Sched*
gamma_c_hihi	8050844.500 [± 600034.582]	7610176.437 [± 697710.369]	**7607296.918** [± **631202.153**]
gamma_c_hilo	156249.200 [± 18836.883]	155251.196 [± 16533.074]	**154812.745** [± **10846.362**]
gamma_c_lohi	258756.770 [± 19323.273]	248466.775 [± 21665.664]	**247899.224** [± 24492.211]
gamma_c_lolo	5272.250 [± 770.943]	5226.972 [± 251.945]	**5210.134** [± **240.110**]
gamma_i_hihi	3104762.500 [± 703298.094]	**3077705.818** [± 1102800.748]	3078056.432 [± 888645.874]
gamma_i_hilo	75816.130 [± 4562.552]	75924.023 [± 4173.978]	**75699.814** [± **3295.872**]
gamma_i_lohi	107500.720 [± 21982.994]	**106069.101** [± **18348.615**]	107342.096 [± 24526.384]
gamma_i_lolo	2614.390 [± 133.731]	**2613.110** [± **149.972**]	2622.514 [± 168.561]
gamma_s_hihi	4566206.000 [± 578839.375]	4359312.628 [± 663325.239]	**4343467.785** [± **853673.523**]
gamma_s_hilo	98519.400 [± 7448.582]	98334.640 [± 4526.485]	**98179.9684** [± **6191.471**]
gamma_s_lohi	130616.530 [± 37307.500]	127641.889 [± 44480.9]	**126822.766** [± **53702.900**]
gamma_s_lolo	3583.440 [± 276.094]	**3515.526** [± **279.368**]	3555.009 [± 218.364]

In the case of makespan optimization, *HGS-Sched* scheduler outperforms the Braunt's GA in all instances, and the Xhafa's GA in 75% of considered instances. The hierarchical algorithm seems to be the most effective for consistent and semi-consistent ETC matrices, and it is worse than single-population Xhafa's struggle

Table 4.7 Average values of *Mean_Flowtime* for benchmark instances [±*s.d.*]: –standard deviation

Instance	Braunt's GA	Xhafa's GA	*HGS-Sched*
gamma_c_hihi	1048333229.742 [± 62551800.953]	1039048563.591 [± 108789000.034]	**1038849002.691 [± 95274900.026]**
gamma_c_hilo	27687019.467 [± 1855790.564]	27620519.758 [± 1572700.532]	**27397760.132 [± 1761150.483]**
gamma_c_lohi	34767197.164 [± 1662420.222]	34566883.883 [± 4060500.544]	**34501187.333 [± 4362160.643]**
gamma_c_lolo	920475.174 [± 51414.611]	917647.316 [± 56050.154]	**915488.420 [± 34862.677]**
gamma_i_hihi	378010732.653 [± 108287000.473]	379768078.537 [± 190414000.845]	**357299609.516 [± 148015000.873]**
gamma_i_hilo	12775104.787 [± 894854.731]	12674329.173 [± 749330.642]	**12517223.618 [± 680981.333]**
gamma_i_lohi	13444708.394 [± 3503680.766]	13417596.793 [± 3324080.433]	**12819181.271 [± 4401610.934]**
gamma_i_lolo	446695.831 [±25385.445]	440728.985 [± 34978.732]	**440511.843 [± 32745.674]**
gamma_s_hihi	526866515.467 [± 105362000.654]	524874694.232 [± 126013000.322]	**521884731.434 [± 134002000.119]**
gamma_s_hilo	16598635.595 [± 1244710.342]	**16372763.267 [± 1078820.565]**	16432276.198 [± 1157260.442]
gamma_s_lohi	15644101.363 [± 4981350.746]	15639622.548 [± 6657510.543]	**15322972.987 [± 8207100.773]**
gamma_s_lolo	605375.388 [± 37980.799]	**598332.695 [± 59455.762]**	602565.625 [± 43452.103]

GA for inconsistent matrices. The efficiency of the hierarchical scheduler is better in the case of flowtime minimization: *HGS-Sched* performs coherently for all three groups of instances – consistent, semi-consistent and inconsistent ETC matrices.

4.4 The Empirical Study in Large-Scale Static and Dynamic Instances

The benchmarks for large-scale static and dynamic grid scenarios were generated by the main module of *Sim-G-Batch* simulator with the input parameters defined in Sec. 4.2. The experiments were performed for two multi-population genetic schedulers, namely *HGS-Sched* and *Island GA*, single-population GA schedulers and a hybrid meta-heuristic defined by GA and *Tabu Search* methods.

Similarly to the previous section, the empirical analysis was conducted in two main steps: the tuning process of the genetic engine for the hybrid and multi-population schedulers, and the comparative analysis of all types of metaheuristics.

4.4.1 The Tuning the GA Engine

The tuning process of GA schedulers in large-scale grids has been provided for the same set of the genetic operators as this specified in Sec. 4.3.2. Eighteen GA variants with all possible combinations of these operators are defined in Table 4.8.

The values of key parameters a general HG-Sched model for each single-population scheduler are shown in Table 4.9. These parameters were tuned empirically in the study provided by the research group of Fatos Xhafa (see [161], [158], [159]) and the previous study of the author of this book [90], [86].

The *HGS-Sched* has in this case just one core branch. Eighteen GA metaheuristics were evaluated in the static and the dynamic grid environments. Each experiment was repeated 30 times under the same configuration of operators and parameters. The averaged $Makespan$ and $Mean\_Flowtime$ values obtained by all schedulers are presented in Tables 4.10–4.13.

Tables 4.15 and 4.15 summarize the relative ranking of eighteen scheduling algorithms for $Makespan$ and $Mean\_Flowtime$ minimization in both the static and the dynamic grid scenarios.

It follows from the results that the quality of scheduling strongly depends on the proper combination of crossover and mutation operations. In all considered instances the *PMX* crossover together with *Move* mutation give the worst results and the combination of CX crossover with *Rebalancing* mutation seems to be the most effective in the most of the instances. Indeed, in the static case $GA-CX-R-ST$ algorithm ranks first in 75% of instances and $GA-CX-R-SS$ algorithm is the best in the remaining 25%of the total instances. In the dynamic scenario the situation is similar: $GA-CX-R-ST$ algorithm achieves the best results in 62.5% of instances and $GA-CX-R-SS$ – in the remaining 27.5%. The $GA-CX-R-EG$ algorithm ranks in all cases as the second or third best scheduler. I can be observed that in the groups of algorithm with the same mutation and crossover operators, the *Struggle* replacement mechanism has the best positive impact on the algorithm performance.

Table 4.8 Eighteen variants of single-population GA-based schedulers

Scheduler	Crossover method	Mutation method	Replacement method
GA-PMX-M-SS	Partially Matched (PMX)	Move	Steady State
GA-PMX-M-EG	Partially Matched (PMX)	Move	Elitist Generational
GA-PMX-M-ST	Partially Matched (PMX)	Move	Struggle
GA-PMX-S-SS	Partially Matched (PMX)	Swap	Steady State
GA-PMX-S-EG	Partially Matched (PMX)	Swap	Elitist Generational
GA-PMX-S-ST	Partially Matched (PMX)	Swap	Struggle
GA-PMX-R-SS	Partially Matched (PMX)	Rebalancing	Steady State
GA-PMX-R-EG	Partially Matched (PMX)	Rebalancing	Elitist Generational
GA-PMX-R-ST	Partially Matched (PMX)	Rebalancing	Struggle
GA-CX-M-SS	Cycle (CX)	Move	Steady State
GA-CX-M-EG	Cycle (CX)	Move	Elitist Generational
GA-CX-M-ST	Cycle (CX)	Move	Struggle
GA-CX-S-SS	Cycle (CX)	Swap	Steady State
GA-CX-S-EG	Cycle (CX)	Swap	Elitist Generational
GA-CX-S-ST	Cycle (CX)	Swap	Struggle
GA-CX-R-SS	Cycle (CX)	Rebalancing	Steady State
GA-CX-R-EG	Cycle (CX)	Rebalancing	Elitist Generational
GA-CX-R-ST	Cycle (CX)	Rebalancing	Struggle

By summarizing the results for both objective functions, the *GA-CX-R-ST* algorithm is selected as the genetic engine for the experimental analysis of multilevel grid scheduling presented in the following section.

Table 4.9 GA setting for large static and dynamic benchmarks

Parameter	Elitist Generational/Struggle	Steady State
degree of branches (t)	0	
period_of_metaepoch (α)	$1/2*n/10$	$(2*n/10$
nb_of_metaepochs	10	
population size (pop_size)	$\lceil(\log_2 n)^2 - \log_2 n\rceil$	$4*(\log_2 n - 1)$
intermediate pop.	$pop\_size - 2$	$(pop\_size)/3$
cross probab.	0.8	1.0
mutation probab.	0.2	
initialization	LJFR-SJFR + MCT + Random	
max_time_to_spend	40 sec. $(static)$ / 75 sec. $(dynamic)$	

4.4.2 The Empirical Evaluation of Single-, Multi-Population and Hybrid Genetic-Based Schedulers in Static and Dynamic Scenarios

The objective of the study presented in this section is to verify and compare the efficiency of single population genetic-based schedulers with multi-level meta-heuristics, namely *HGS-Sched* and *GA-CX-R-SS* hybridized with Tabu Search method. Based on the evaluation results of all GA variants we implemented the *GA-CX-R-SS* as a basic genetic mechanism in *HGS-Sched*.

4.4.2.1 Metaheuristics for Study

The following single- and multi-population genetic schedulers are considered in this part of the empirical analysis:

- *GA-CX-R-ST* - a single population genetic algorithm generated as the best scheduler in the previous section (Sec. 4.4.1) with *CX* crossover, *Rebalancing* mutation and *Struggle* replacement operators. This algorithm is used as the main genetic mechanism in the remaining two multi-level techniques and hybrid strategy. The $GA-CX-R-ST$ settings are given in Table 4.9.
- *HGS-Sched* with $GA-CX-R-ST$ with various population sizes and mutation rates in the branches;
- *IGA* - Island Genetic Algorithm with $GA-CX-R-ST$ as the basic mechanism in all sub-populations;

Table 4.10 Average *Makespan* values for eighteen GA-based schedulers in static instances [$\pm s.d.$], ($s.d.$ = standard deviation)

Strategy	Small	Medium	Large	Very Large
GA-PMX-M-SS	4162673.327 [± 790668.957]	4298014.024 [± 824291.171]	4484782.436 [± 911521.151]	4464900.721 [± 560336.505]
GA-PMX-M-EG	4178242.873 [± 209829.920]	4300135.827 [± 552350.473]	4480981.012 [± 375819.451]	4486359.298 [± 750553.881]
GA-PMX-M-ST	4157240.307 [± 379850.512]	4295825.090 [± 415189.756]	4477518.527 [± 449752.022]	443533.726 [± 364469.067]
GA-PMX-S-SS	4125691.126 [± 659448.392]	4268348.675 [± 475553.050]	4401693.348 [± 673682.970]	4417588.754 [± 853210.688]
GA-PMX-S-EG	4148405.275 [± 43761.015]	4285671.797 [± 605962.237]	4416301.090 [± 53486.153]	4421427.663 [± 701992.116]
GA-PMX-S-ST	4133822.183 [± 380836.642]	4261593.241 [± 634705.248]	4399261.431 [± 555304.633]	439189.653 [± 711735.974]
GA-PMX-R-SS	4099722.422 [± 939509.617]	4209834.534 [± 505720.705]	4322903.467 [± 149437.882]	4332736.028 [± 563164.075]
GA-PMX-R-EG	4111018.744 [± 837452.949]	4217551.633 [± 540610.017]	4361359.893 [± 590881.527]	4375643.075 [± 748754.806]
GA-PMX-R-ST	4096457.271 [± 553704.095]	4202740,309 [± 468878.756]	4309984.381 [± 497267.784]	4313319.068 [± 552852.618]
GA-CX-M-SS	4057867.000 [± 711772.879]	4175408.001 [± 796525.118]	4234538.827 [± 933600.720]	4284402.774 [± 684131.232]
GA-CX-M-EG	4067320.903 [± 654607.977]	4184339.982 [± 582147.023]	4290330.342 [± 751364.567]	4303769.532 [± 952170.387]
GA-CX-M-ST	4041129.964 [± 987767.110]	4167727.904 [± 803611.007]	4209105.106 [± 679072.949]	4241630.898 [± 728165.507]
GA-CX-S-SS	3954447.328 [± 566725.393]	4125317.265 [± 771196.560]	4172916.118 [± 618157.753]	4205709.695 [± 944696.828]
GA-CX-S-EG	4004965.627 [± 918998.520]	4156258.860 [± 713229.061]	4190734.257 [± 594713.746]	4222631.851 [± 957740.279]
GA-CX-S-ST	3940188.666 [± 399958.718]	4098993.759 [± 775636.448]	4161474.891 [± 65051.579]	4219927.630 [± 828429.144]
GA-CX-R-SS	3923847.672 [± 384195.770]	4053438.673 [± 697330.842]	4105672.240 [± 581510.738]	4152951.283 [± 712805.158]
GA-CX-R-EG	3912183.019 [± 611208.560]	4095293.387 [± 621501.898]	4093277.276 [± 984994.621]	4197426.188 [± 584731.707]
GA-CX-R-ST	**3907275.654 [± 472756.651]**	**3982875.764 [± 527598.824]**	**4065433.887 [± 628860.552]**	**4100554.634 [± 684458.248]**

Table 4.11 Average *Makespan* values for eighteen GA-based schedulers in dynamic instances [$\pm s.d.$], ($s.d.$ = standard deviation)

Strategy	Small	Medium	Large	Very Large
GA-PMX-M-SS	4183064.389 [± 303594.681]	41996544.500 [± 633541.835]	4272342.027 [± 523861.891]	4351579.014 [± 955889.372]
GA-PMX-M-EG	4196660.637 [± 795391.867]	4207559.346 [± 673312.285]	4250769.729 [± 570063.523]	4378943.912 [± 663276.562]
GA-PMX-M-ST	41666682.922 [± 557522.757]	4182062.259 [± 659041.716]	4242506.997 [± 564267.992]	4342520.961 [± 435640.377]
GA-PMX-S-SS	4150255.659 [± 364204.730]	4156125.373 [± 612692.293]	4239582.165 [± 566626.175]	4323745.969 [± 685834.282]
GA-PMX-S-EG	4149105.038 [± 440578.157]	416379.853 [± 472235.032]	4220166.253 [± 837511.681]	4309346.443 [± 515978.598]
GA-PMX-S-ST	4133090.502 [± 271571.051]	4177107.429 [± 473557.935]	4198308.287 [± 492853.075]	4294855.678 [± 430583.789]
GA-PMX-R-SS	4126676.780 [± 728966.192]	4118808.678 [± 649630.565]	4164756.208 [± 56766.709]	4254796.820 [± 580607.466]
GA-PMX-R-EG	4129602.691 [± 270662.077]	4109778.902 [± 913428.430]	4182790.083 [± 761523.938]	4270973.780 [± 100219.228]
GA-PMX-R-ST	4063022.741 [± 430715.830]	4077928.331 [± 410172.944]	4158207.631 [± 611305.330]	4232768.084 [± 499217.899]
GA-CX-M-SS	4096232.073 [± 554669.949]	4093218.110 [± 602133.299]	4114422.323 [± 721311.202]	4188744.263 [± 672746.995]
GA-CX-M-EG	4109712.181 [± 869717.384]	4099788.995 [± 453095.636]	4141965.814 [± 817008.729]	4209289.242 [± 636853.293]
GA-CX-M-ST	4054429.908 [± 161589.755]	4070157.151 [± 116128.587]	4127500.870 [± 870899.646]	4198327.481 [± 631520.079]
GA-CX-S-SS	4022572.196 [± 516330.313]	4032408.549 [± 590815.709]	4090392.967 [± 898536.995]	4129281.695 [± 917216.254]
GA-CX-S-EG	4015442.255 [± 405905.662]	4033181.392 [± 650601.916]	4106452.250 [± 982489.800]	4164853.468 [± 785931.696]
GA-CX-S-ST	4006934.189 [± 436482.946]	4010678.953 [± 516992.488]	4039201.986 [± 816514.067]	4140525.379 [± 633513.047]
GA-CX-R-SS	3986872.198 [± 806632.171]	3972483.354 [± 618035.814]	**4024651.986 [± 436431.848]**	4109893.365 [± 747400.818]
GA-CX-R-EG	3998111.691 [± 692024.001]	4002987.983 [± 467779.063]	4089790.044 [± 632965.292]	4091857.920 [± 791598.934]
GA-CX-R-ST	**3956624.255 [± 521037.286]**	**3968890.040 [± 659955.935]**	4033653.781 [± 600996.138]	**4040335.820 [± 642477.287]**

Table 4.12 Average *Mean_Flowtime* values for eighteen GA-based schedulers in static instances [$\pm s.d.$], ($s.d.$ = standard deviation)

Strategy	Small	Medium	Large	Very Large
GA-PMX-M-SS	1193567211.537 [± 128370293.469]	2309597733.454 [± 291748525.988]	4399334723.599 [± 411586354.582]	8401592320.869 [± 889213844.016]
GA-PMX-M-EG	1206734272.356 [± 135504556.429]	2326428289.314 [± 13744410.910]	4404322051.418 [± 439654339.019]	8410389312.587 [± 989122999.052]
GA-PMX-M-ST	1182232499.779 [± 218224825.514]	2297351808.190 [± 207359074.643]	4357752153.775 [± 476389087.741]	8395922739.597 [± 606291276.896]
GA-PMX-S-SS	1158140985.025 [± 180302378.918]	2261813008.415 [± 240332696.510]	4316860418.277 [± 593778909.154]	8367766411.152 [± 755794904.240]
GA-PMX-S-EG	1175542598.454 [± 162987312.683]	228568131.930 [± 369229026.877]	4332618330.048 [± 45367872.253]	8388110055.663 [± 507151151.344]
GA-PMX-S-ST	1147247760.724 [± 211590465.457]	2271025623.997 [± 453137510.918]	4320515589.851 [± 711100452.764]	8373206007.045 [± 784801239.205]
GA-PMX-R-SS	1142225326.145 [± 240067553.885]	2249643747.642 [± 253050662.960]	4286055243.299 [± 577523722.350]	8338732539.636 [± 741000998.450]
GA-PMX-R-EG	1136071183.723 [± 143892082.090]	2258614783.371 [± 233840747.951]	4292713481.576 [± 377833946.994]	835216627.023 [± 699126484.932]
GA-PMX-R-ST	1117878614.800 [± 196255094.628]	2227564670.521 [± 225294411.337]	4283126157.824 [± 543390178.534]	8322161405.019 [± 736033399.656]
GA-CX-M-SS	1105676446.564 [± 18596758.737]	2195359732.256 [± 282655389.940]	4263563557.141 [± 539894530.830]	8294397268.110 [± 790327708.149]
GA-CX-M-EG	1125655077.793 [± 142293067.762]	2238431289.893 [± 212224990.911]	4274715011.673 [± 604531703.308]	8306547838.652 [± 812904982.726]
GA-CX-M-ST	11111848829.461 [± 110335889.154]	2204221008.299 [± 254605827.121]	4238467253.243 [± 513415557.850]	8272449832.295 [± 822664629.425]
GA-CX-S-SS	1084338418.886 [± 165598016.948]	2179625719.400 [± 250682728.242]	4204403080.180 [± 430849430.300]	8242964769.102 [± 600614818.712]
GA-CX-S-EG	1097868914.479 [± 17614361.064]	2190763974.690 [± 334857074.927]	4227164271.644 [± 473657175.059]	8260315502.581 [± 569231789.391]
GA-CX-S-ST	1078132035.030 [± 120473769.993]	2166781141.473 [± 250660461.860]	4195675260.221 [± 577482551.646]	8253290991.790 [± 769767684.495]
GA-CX-R-SS	1060575340.426 [± 123314394.677]	2138208217.698 [± 362885985.554]	**4188077792.436 [± 537308748.955]**	**8213447951.179 [± 803914314.541]**
GA-CX-R-EG	1066071183.009 [± 103578848.015]	2143852396.768 [± 272146853.672]	4196929745.169 [± 533701714.987]	8235408309.560 [± 795752579.317]
GA-CX-R-ST	**1054421614.528 [± 15296382.605]**	**2111081930.859 [± 339791854.177]**	4190750381.697 [± 436450229.143]	8221050176.578 [± 708270871.387]

Table 4.13 Average *Mean_Flowtime* values for eighteen GA-based schedulers in dynamic instances [±*s.d.*], (*s.d.* = standard deviation)

Strategy	Small	Medium	Large	Very Large
GA-PMX-M-SS	1343724728.245 [± 598538835.360]	2321846250.347 [± 431346593.814]	4413245472.632 [± 417986755.794]	8644534678.245 [± 404482983.284]
GA-PMX-M-EG	1389239424.349 [± 129097532.581]	2347268532.324 [± 432955978.981]	4436061767.548 [± 701148099.631]	8660435684.376 [± 664054585.165]
GA-PMX-M-ST	1307355142.051 [± 126906586.696]	2308727439.878 [± 256156869.233]	4403766189.879 [± 469026059.421]	8605175251.450 [± 693219859.321]
GA-PMX-S-SS	1263649999.867 [± 268317752.960]	2262608448.916 [± 216787358.248]	4360160544.425 [± 656197729.295]	8502620979.464 [± 851502055.485]
GA-PMX-S-EG	1286784630.430 [± 184783757.268]	2292058041.397 [± 499461667.383]	4393131389.346 [± 454805534.425]	8531596508.841 [± 573812983.895]
GA-PMX-S-ST	1237247760.724 [± 244213856.135]	2271025623.997 [± 434633012.674]	4377515589.851 [± 788991696.341]	8513206007.045 [± 719513808.715]
GA-PMX-R-SS	1226429310.580 [± 255806730.308]	2229357685.290 [± 290693183.985]	4333974209.902 [± 519849125.536]	8472673073.563 [± 743945636.397]
GA-PMX-R-EG	1213613356.567 [± 167571940.967]	2254728732.731 [± 283645546.676]	4348829866.347 [± 526061100.837]	8493687750.198 [± 784339554.900]
GA-PMX-R-ST	1151930817.794 [± 134663685.381]	2184860096.988 [± 242728273.618]	4320407124.381 [± 528711756.114]	8466099314.428 [± 728490457.382]
GA-CX-M-SS	1188284638.934 [± 162112136.850]	2199577483.835 [± 257276315.811]	4239455642.778 [± 387064110.517]	8426829568.357 [± 537095310.247]
GA-CX-M-EG	1193593276.050 [± 163978670.974]	2211445673.176 [± 173050142.794]	4296573568.612 [± 621630992.871]	8457381627.647 [± 596961998.050]
GA-CX-M-ST	1132978230.586 [± 129547599.824]	2166440897.331 [± 182827503.146]	4270646610.327 [± 616490962.213]	8432573052.527 [± 885691236.077]
GA-CX-S-SS	1126523158.454 [± 228154433.283]	2161517451.690 [± 277432894.907]	4222289600.939 [± 552469744.108]	8365526025.239 [± 773215992.243]
GA-CX-S-EG	1113772162.066 [± 187159613.747]	2154839647.617 [± 326230384.931]	4233669538.605 [± 620551577.626]	8392873916.020 [± 859689519.119]
GA-CX-S-ST	1106340445.440 [± 138247946.733]	2146356979.846 [± 239873730.726]	4218990182.691 [± 525735729.121]	8345528446.458 [± 947608913.414]
GA-CX-R-SS	1076534164.177 [± 181982147.103]	**2116783653.774 [± 19799961.957]**	421654378.495 [± 48988254.993]	**8291108455.913 [± 632222716.240]**
GA-CX-R-EG	1098943746.287 [± 143403467.472]	2138965387.563 [± 176627923.813]	4208567534.205 [± 564619337.921]	8339386800.606 [± 730340037.273]
GA-CX-R-ST	**1028717087.273 [± 177645283.759]**	2122230062.891 [± 217213422.096]	**4178971901.269 [± 528146731.361]**	8310330051.728 [± 711071173.232]

Table 4.14 Ranking of eighteen genetic-based schedulers in static instances

Strategy	Makespan				Flowtime			
	Small	Medium	Large	Very Large	Small	Medium	Large	Very Large
GA-PMX-M-SS	17	17	18	17	17	17	17	17
GA-PMX-M-EG	18	18	17	18	18	18	18	18
GA-PMX-M-ST	16	16	16	16	16	16	16	16
GA-PMX-S-SS	13	14	14	14	14	13	13	13
GA-PMX-S-EG	15	15	15	15	15	15	15	15
GA-PMX-S-ST	14	13	13	13	13	14	14	14
GA-PMX-R-SS	11	11	11	11	12	11	11	11
GA-PMX-R-EG	12	12	12	12	11	12	12	12
GA-PMX-R-ST	10	10	10	10	9	9	10	10
GA-CX-M-SS	8	8	8	8	7	7	8	8
GA-CX-M-EG	9	9	9	9	10	10	9	9
GA-CX-M-ST	7	7	7	7	8	8	7	7
GA-CX-S-SS	5	5	5	4	5	5	5	4
GA-CX-S-EG	6	6	6	6	6	6	6	6
GA-CX-S-ST	4	4	4	5	4	4	4	5
GA-CX-R-SS	3	2	3	2	2	2	**1**	**1**
GA-CX-R-EG	2	3	2	3	3	3	3	3
GA-CX-R-ST	**1**	**1**	**1**	**1**	**1**	**1**	2	2

- *GA+TS* - hybrid scheduler with $GA - CX - R - ST$ as the control strategy and Tabu Search (TS).

The *Island Genetic Algorithm (IGA)* [153] is a well-known parallel GA technique. An initial (usually big) population is divided into several sub-populations, 'islands' or 'demes', for which single-population GAs with identical configurations of the parameters and operators are activated (one algorithm for each deme). After a fixed number of iterations, denoted as it_d, the migration procedure is activated. It enables a partial exchange (usually according to the standard ring topology) of the individuals among islands. The relative amount of the migrating individuals, represented by

Table 4.15 Ranking of eighteen genetic-based schedulers in dynamic instances

Strategy	Makespan				Flowtime			
	Small	Medium	Large	Very Large	Small	Medium	Large	Very Large
GA-PMX-M-SS	17	17	18	17	17	17	17	17
GA-PMX-M-EG	18	18	17	18	18	18	18	18
GA-PMX-M-ST	16	16	16	16	16	16	16	16
GA-PMX-S-SS	15	13	15	15	14	13	13	13
GA-PMX-S-EG	14	14	14	14	15	15	15	15
GA-PMX-S-ST	13	15	13	13	13	14	14	14
GA-PMX-R-SS	11	12	11	11	12	11	11	11
GA-PMX-R-EG	12	11	12	12	11	12	12	12
GA-PMX-R-ST	8	8	10	10	8	8	10	10
GA-CX-M-SS	9	9	7	7	9	9	7	7
GA-CX-M-EG	10	10	9	9	10	10	9	9
GA-CX-M-ST	7	7	8	8	7	7	8	8
GA-CX-S-SS	6	5	5	4	6	6	5	5
GA-CX-S-EG	5	6	6	6	5	5	6	6
GA-CX-S-ST	4	4	3	5	4	4	4	4
GA-CX-R-SS	2	2	**1**	3	2	**1**	3	**1**
GA-CX-R-EG	3	3	4	2	3	3	2	3
GA-CX-R-ST	**1**	**1**	2	**1**	**1**	2	**1**	2

mig, is the algorithm global parameter commonly known as the *migration rate*, and calculated as:

$$mig = \frac{m_{deme}}{deme} \cdot 100\% \tag{4.5}$$

where *deme* is the size of the sub-population in *IGA* and m_{deme} is the number of migrating individuals in each deme.

The general concept of the 'islands' and 'migration' procedure in *IGA* is presented in Fig. 4.1.

The procedure of migration is repeated after each execution of it_d iterations of genetic algorithm in each sub-population.

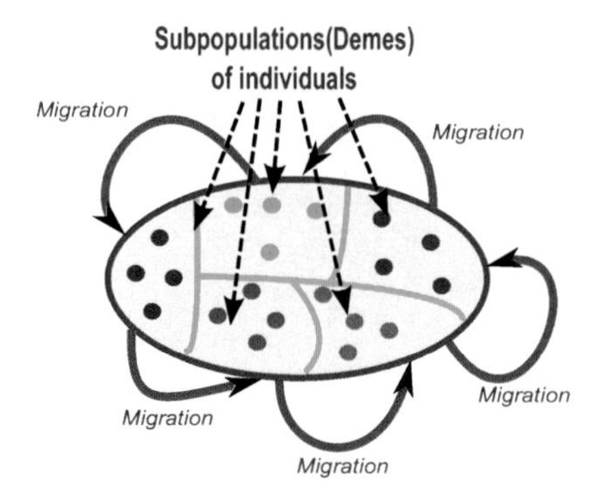

Fig. 4.1 A concept of Island GA

The general flow of the $GA + TS$ hybrid scheduler is presented in Fig. 4.2.

Tabu Search (TS) algorithm is implemented as subordinate module in the GA scheduler for the replacement of the mutation operation in $GA - CX - R - ST$, i.e. selection, crossover and mutation in GA is replaced by selection, crossover and TS procedures. In each iteration of the $GA + TS$ algorithm, the TS method starts from the population generated as the result of application of the crossover procedure in $GA - CX - R - ST$. The structures of chromosomes are modified by 'moving' the current schedules to their neighbors in order to improve their fitness values.

One of the most important features of TS is the use of a *historical memory*, which consists of a *short term memory* (or *recency*) with information on recently visited solutions, and a *long term memory* (or *frequency*), where all information gathered in the exploration of the search space process by TS module is cumulated. Both 'memory' modules are represented as the lists in the TS implementation. The 'movements' in these two lists are considered as 'tabu operations' and then they cannot be activated. Therefore, some *aspiration criteria* are needed for removing the tabu movements. Additionally, some *local search* inner heuristics are also needed for an exploration of a neighborhood of a given solution, and the *intensification and diversification procedures* are necessary for the management of the exploration/exploitation tradeoff in the global search.

The following configuration of TS algorithm is used in this study:

- ***historical memory*** – both short and long term memories are used. For the *recency* memory, a tabu list matrix $TL_{(n \times m)}$ is generated to maintain the tabu status. In addition, a tabu hash table (TH) is used in order to further filter the tabu solutions [164];
- ***neighborhood exploration*** – steepest descent/mildest ascent method with *swap* moving strategy, which exchanges two tasks assigned to different machines [145];
- ***aspiration criterium***– GA fitness function;
- ***intensification*** – this procedure is executed for the detailed exploration of promising regions in the optimization landscape by rewarding (attributes of) the current solution. The structure of the neighborhood of a given solution is modified by

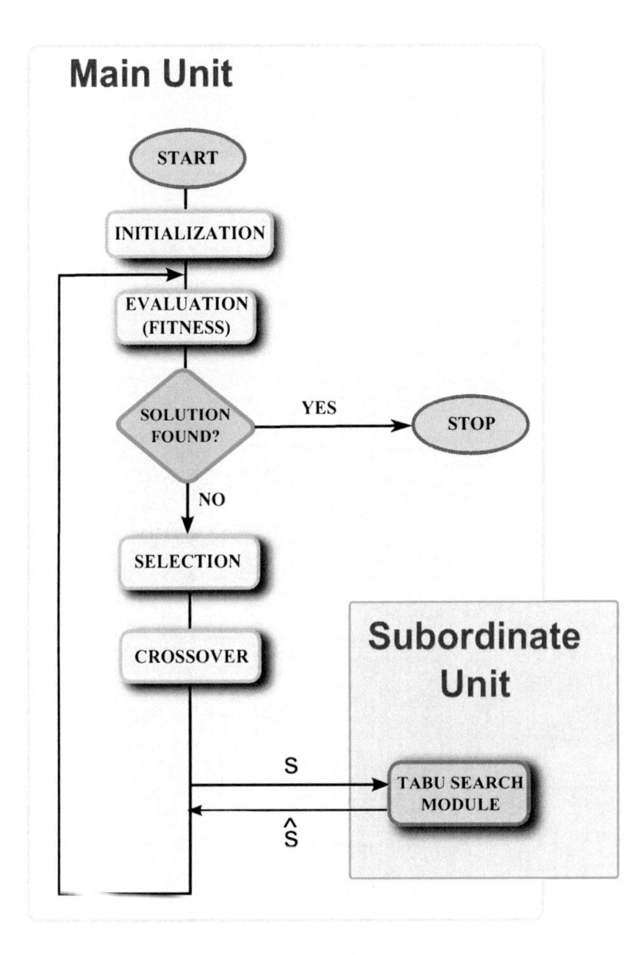

Fig. 4.2 The general flowchart of hybrid GA+TS algorithm

performing all possible swap movements between two machines and just one transfer from one machine to the other [53].

- ***diversification*** – a *soft* diversification method is realized by using a *penalization of ETC values* technique [164], and *strong* diversification method is realized by random modifications in the assignments of a sufficient number of tasks.

The key parameters for $HGS-Sched, IGA$ and $GA+TS$ schedulers used in this part of the empirical study are presented in Tables 4.16, 4.17, and 4.18.

It is assumed that *HGS-Sched* tree is composed of the branches of degrees 0 and 1, and the search process in the sprouted branches is approximately two times slower than in the core (the mutation rates in the sprouted branches are two times lower than the mutation rate in the core).

Table 4.16 *HGS-Sched* settings for static and dynamic benchmarks

Parameter	
$period\_of\_metaepoch$	$20 * n$
$nb\_of\_metaepochs$	10
degrees of branches (t)	0 and 1
population size in the core	$3 * (\lceil 4 * (\log_2 n - 1)/(11.8) \rceil)$
population size in the sprouted branches (b_pop_size)	$(\lceil (4 * (\log_2 n - 1)/(11.8) \rceil)$
intermediate pop. in the core	$abs((r\_pop\_size)/3)$
intermediate pop. in the sprouted branch	$abs((b\_pop\_size)/3)$
cross probab.	0.9
mutation probab. in core	0.4
mutation probab. in the sprouted branches	0.2
$max\_time\_to\_spend$	40 sec. $(static)$ / 70 sec. $(dynamic)$

Table 4.17 Configuration of *IGA* algorithm

Parameter	
it_d	$20 * n$
mig	5 %
number of islands (demes)	10
cross probab.	1.0
mutation probab.	0.2
$max\_time\_to\_spend$	40 sec. $(static)$ / 70 sec. $(dynamic)$

Table 4.18 Configuration of TS algorithm

Parameter	
start_choice	Min-Min method
tabu_size	918133
max_tabu_status	32
max_repetitions	69
nb_diversifications	8
nb_intensifications	8
nb_iterations	8192
elite_size	30
aspiration_value	20
max_time_to_spend	100 seconds

4.4.2.2 Results

Similarly to the GAs tuning analysis, each experiment was repeated 30 times under the same configuration of operators and parameters. The box-plots[4] for the values of the *Makespan* and *Mean_Flowtime* achieved by four considered schedulers are presented in Fig. 4.3–4.10.

It can be observed that in the case of *Makespan* minimization three multi-level and multi-population schedulers outperform the $GA - CX - R - ST$ method in all considered instances, but the differences in the results of these three schedulers are not so significant. *GA+TS* hybrid scheduler is the most efficient in 'Small' static and dynamic grids and the case of 'Very large' static grid. It may stem from the fact that Tabu Search technique is usually accurate in exploration of the highly parametrized local static neighborhoods of the partial solutions found by the genetic steering module. However, this technique may be sensitive to any modifications of the system state. *TS* meta-heuristic cannot guarantee the fast escape from a basin of attraction of already detected local optima in the case of appearance of new promising solutions. Similarly, the island algorithm can explore effectively the search space only in the case of minor changes in the system states. It works quite well in most of the

[4] All statistical analysis was provided and all box-plot charts ware plotted by using the 'STATISTICA PL 9.0' Software Package (®StatSoft, www.statsoft.pl).

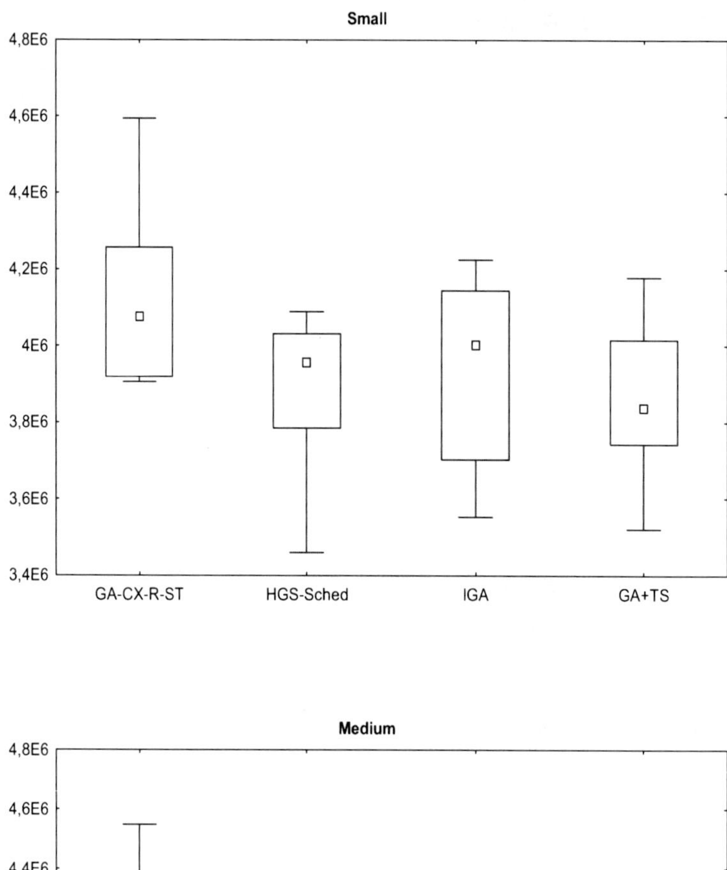

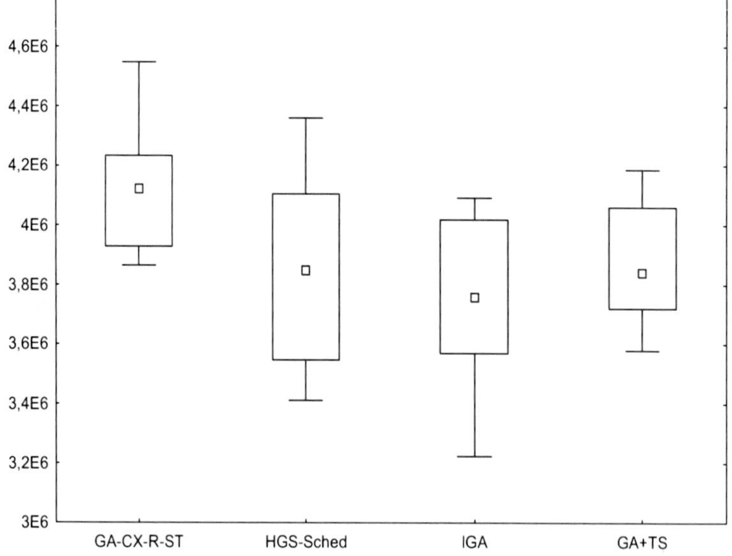

Fig. 4.3 The box-plot of the results for *Makespan* in static scheduling: Small and Medium grids

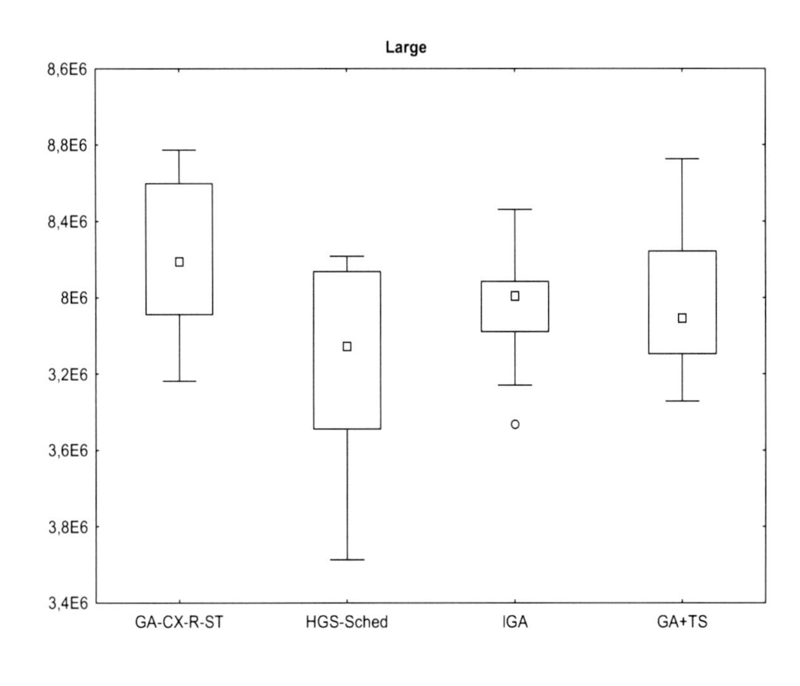

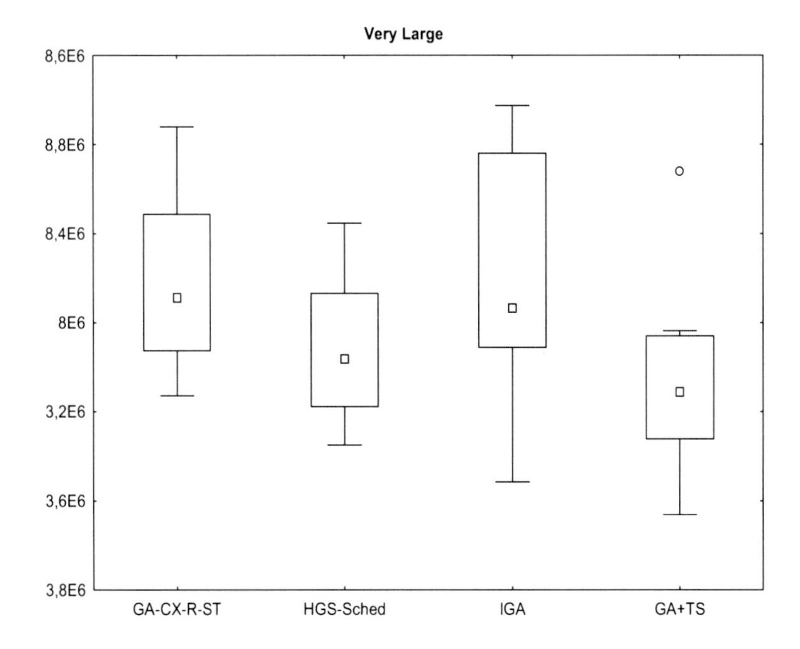

Fig. 4.4 The box-plot of the results for *Makespan* in static scheduling: Large and Very Large grids

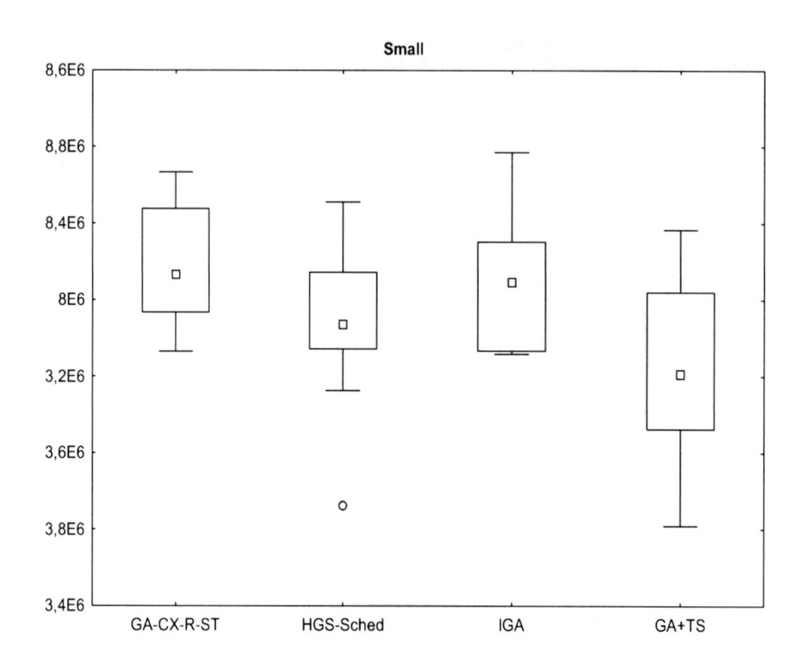

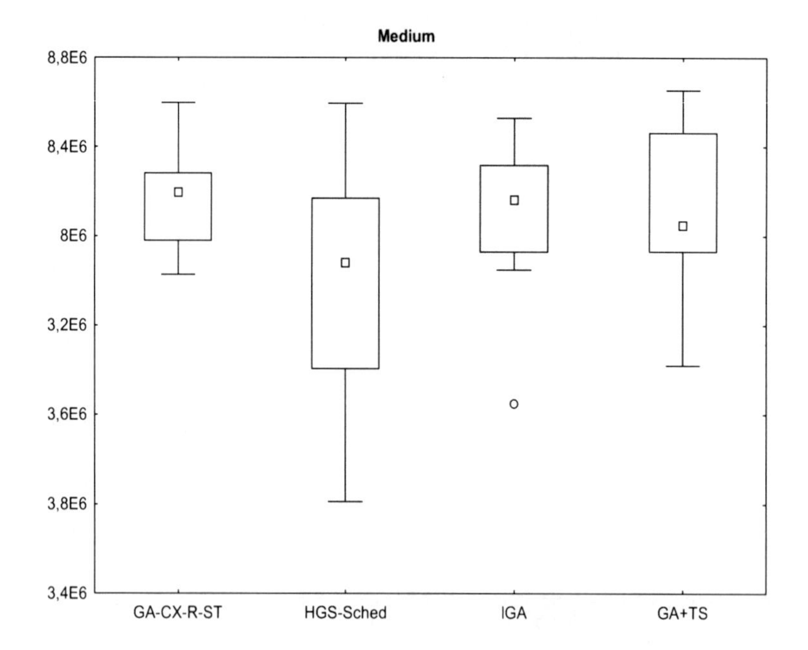

Fig. 4.5 The box-plot of the results for *Makespan* in dynamic scheduling: Small and Medium grids

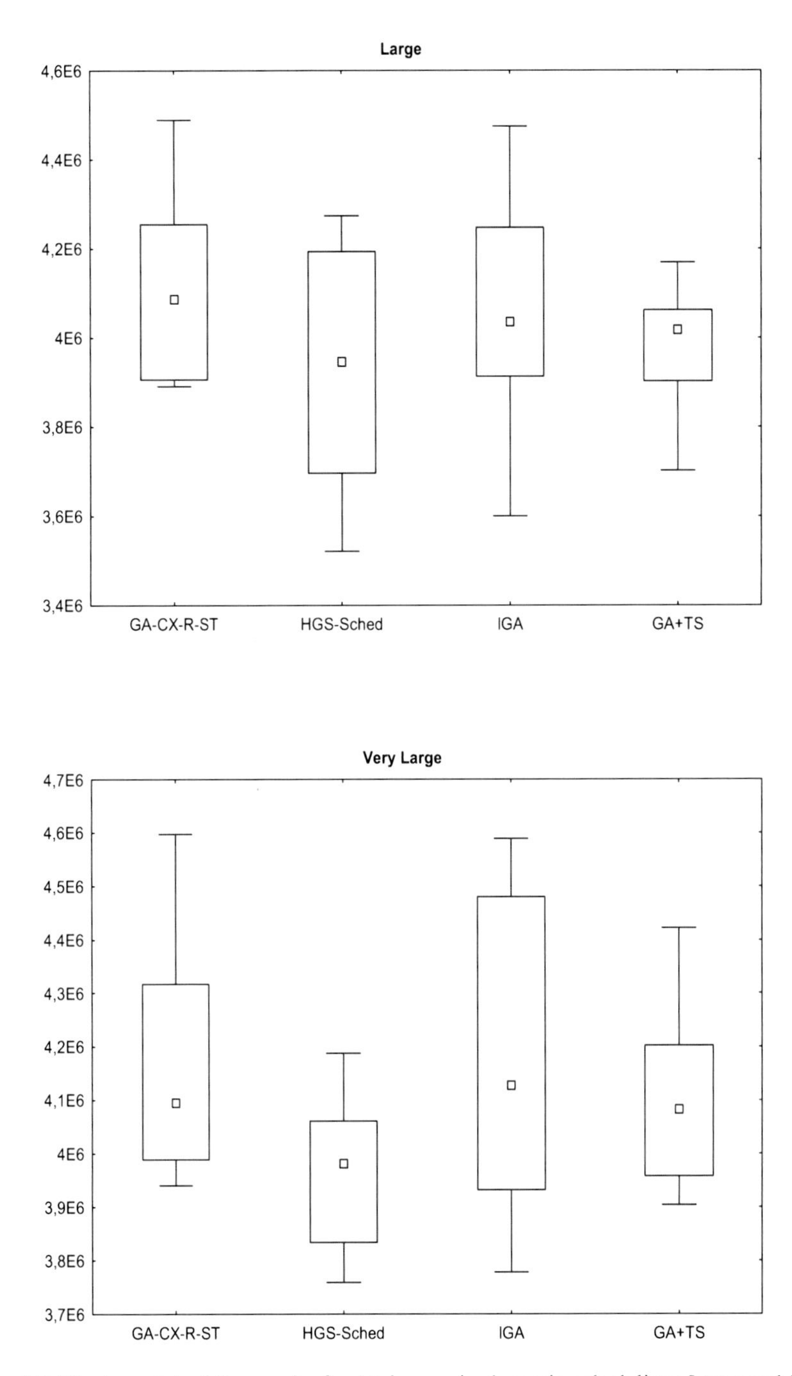

Fig. 4.6 The box-plot of the results for *Makespan* in dynamic scheduling: Large and Very Large grids

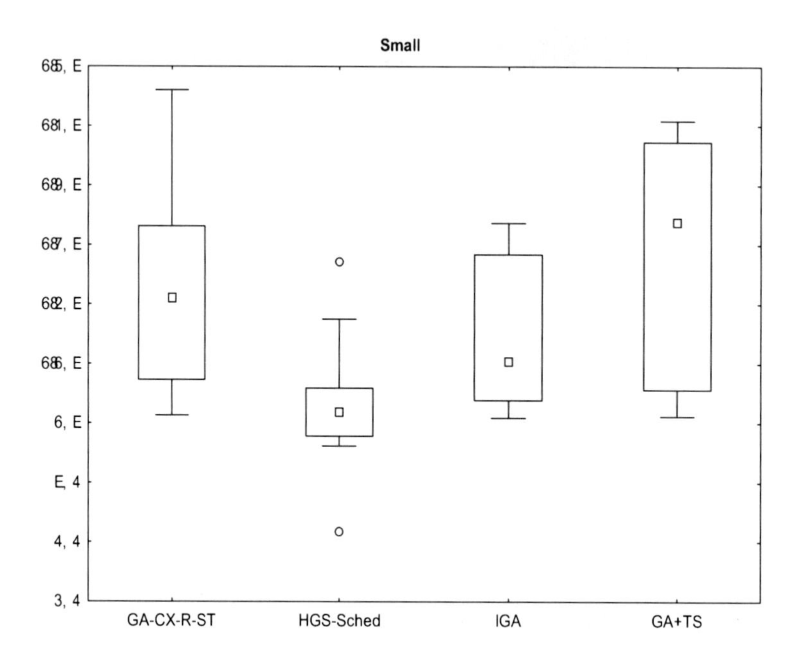

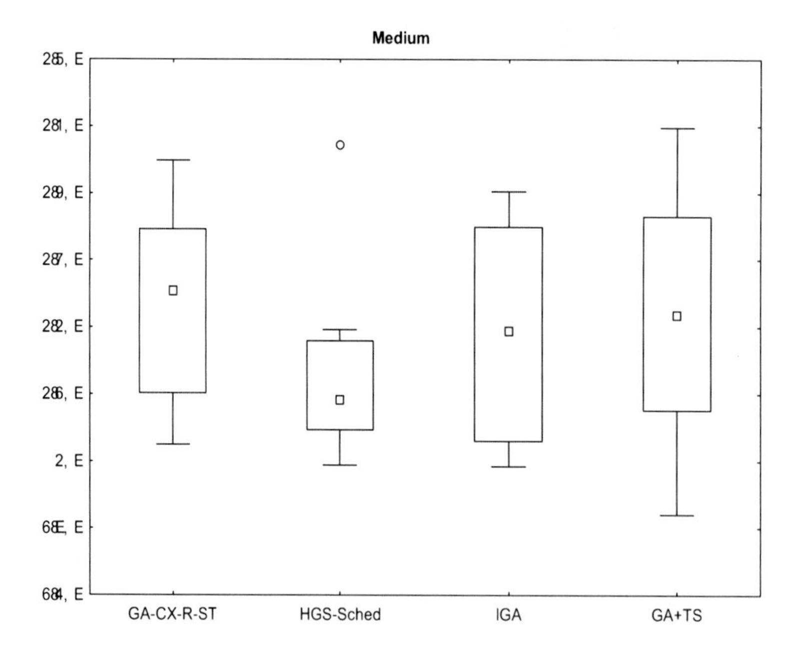

Fig. 4.7 The box-plot of the results for *Mean_Folwtime* in static scheduling: Small and Medium grids

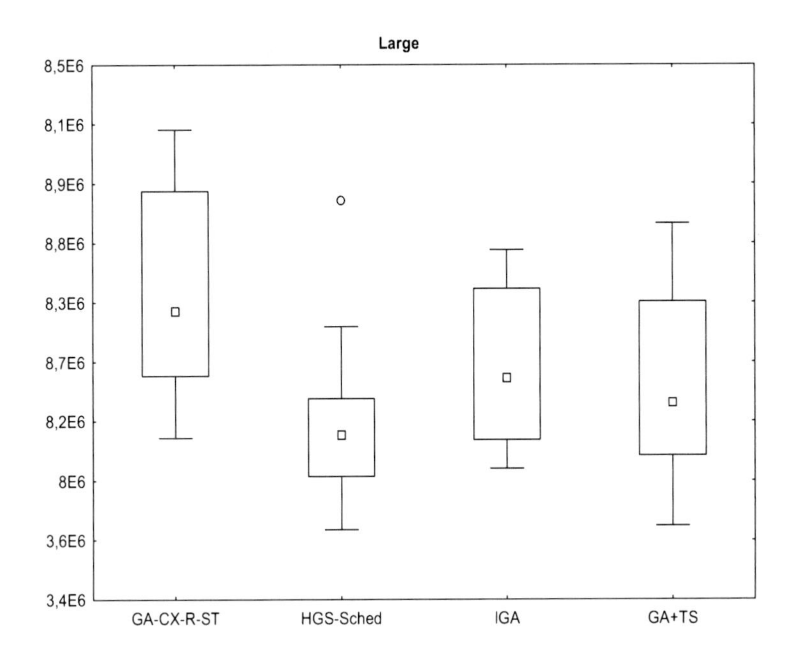

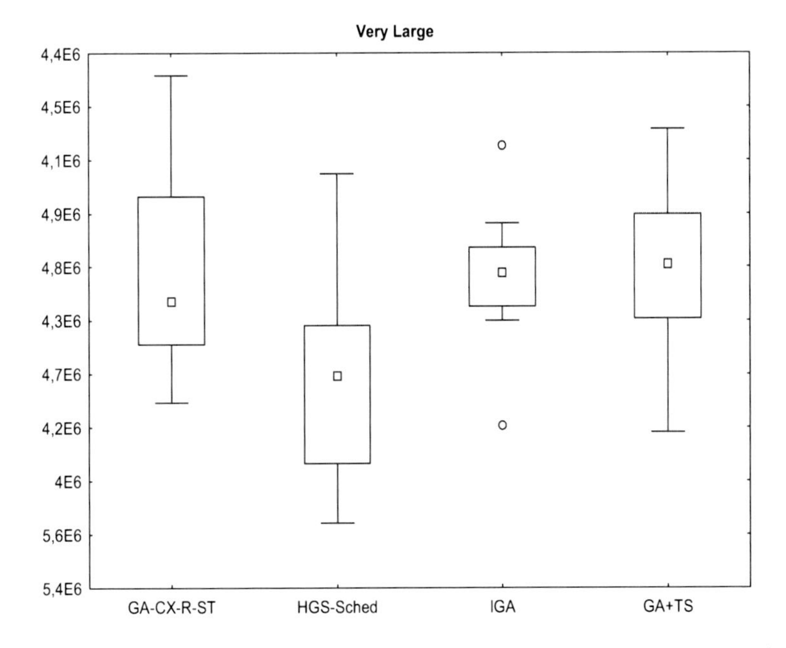

Fig. 4.8 The box-plot of the results for *Mean_Folwtime* in static scheduling: Large and Very Large grids

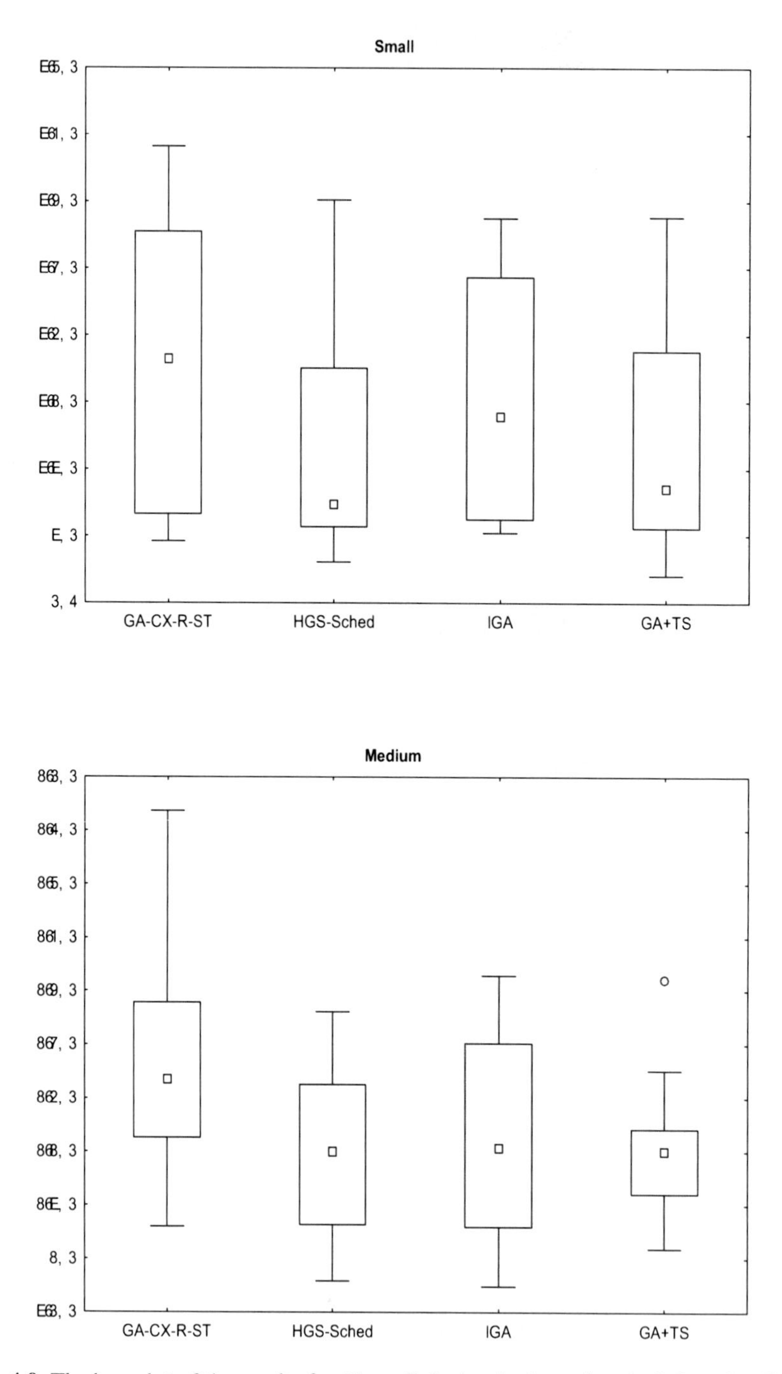

Fig. 4.9 The box-plot of the results for *Mean_Folwtime* in dynamic scheduling: Small and Medium grids

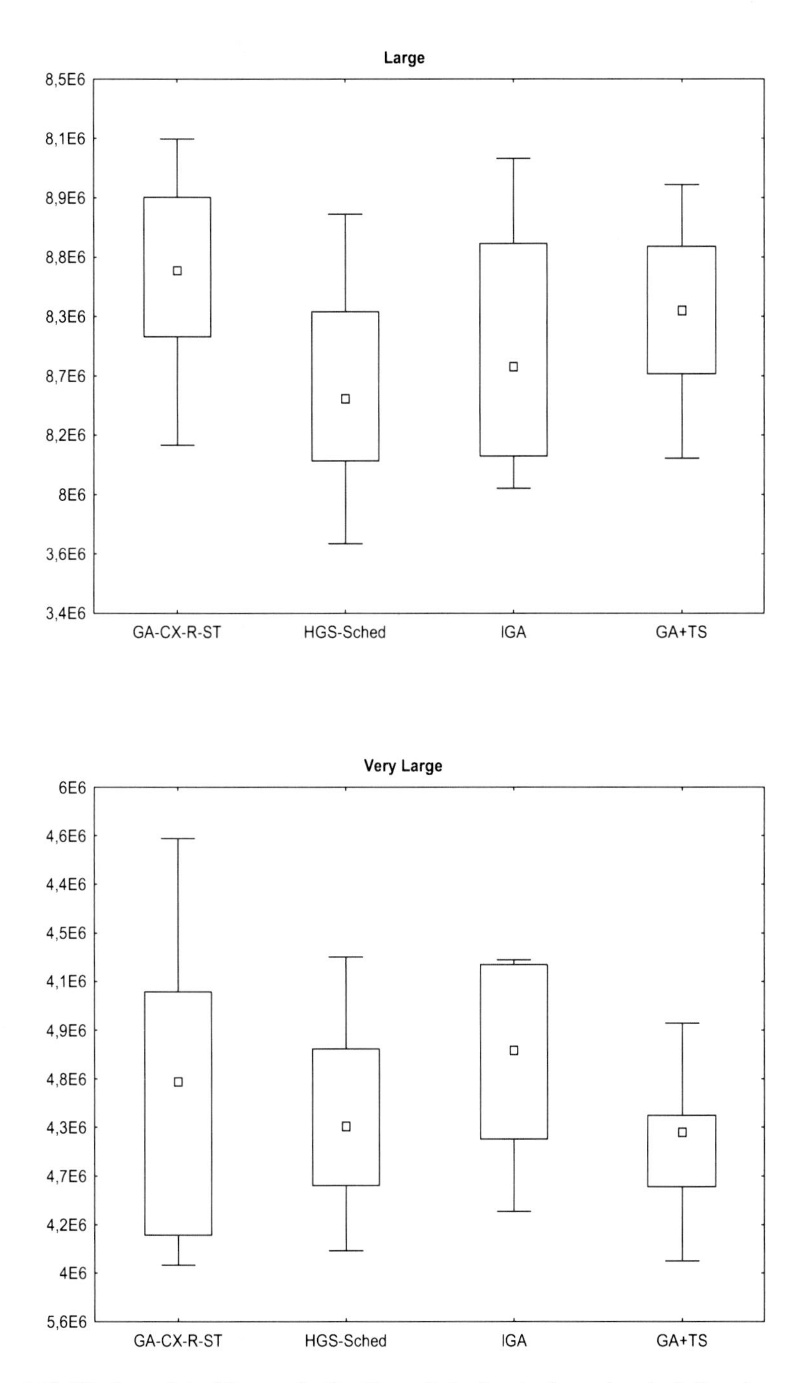

Fig. 4.10 The box-plot of the results for *Mean_Folwtime* in dynamic scheduling: Large and Very Large grids

static scenarios, however the differences between the first and the third quantiles for *IGA* are the largest when compared to the other schedulers. *HGS-Sched* performs very well in most of the dynamic cases.

In the case of *Mean_Flowtime*, hierarchical scheduler outperforms all other meta-heuristics in all instances but 'Very Large' grid in the dynamic scenario. However in this case the difference between the results generated by *HGS-Sched* and *GA+TS* are very minor. It can be observed that the stability of hierarchical scheduler is much better than the stability of the other methods, for which the 'shapes' of the 'boxes' in the plots are relatively big. Those results confirm the high of *HGS-Sched* in the reduction of overall execution time of the schedules, which makes this method well suited to an effective exploration of the new regions in the scheduling space.

A simple comparison of the averaged values of the scheduling objective functions and the standard deviations is usually insufficient to measure the relative performance of the schedulers. Therefore, the standard Student's *t–test* for the comparison of two means [101] was used for the verification of statistical significance of all empirical results of minimization of *Makespan* and *Mean_Flowtime*. The result of this test is the acceptance or rejection of the null hypothesis ($H0$), which states that any differences in results are purely random. An erroneous rejection of the null hypothesis constitutes a ***Type 1*** error.

The best achieved *Makespan* and *Mean_Flowtime* values in each problem instance have been defined as the reference values in the verification of the 'null' hypothesis, and the confidence level was 95 % in each instance of the *t-test*.

Table 4.19 Comparison of the two-tailed *P-values* for *Makespan* and *Mean_Flowtime* results in the large size static and dynamic instances

Strategy	***Makespan***				***Mean_Flowtime***			
	Small	**Medium**	**Large**	**Very Large**	**Small**	**Medium**	**Large**	**Very Large**
					Large Static Instances			
GA-CX-R-ST	0.059	0.185	0.126	0.075	0.119	0.128	0.211	0.055
HGS-Sched	**< 0.0001**	**< 0.0001**	1	**0.031**	1	1	1	1
IGA	**0.012**	1	0.063	0.068	0.072	0.215	0.051	0.185
GA+TS	1	**< 0.0001**	0.058	1	0.065	0.098	**0.049**	0.115
					Dynamic Instances			
GA-CX-R-ST	**0.049**	0.055	**0.049**	**0.042**	0.112	0.057	0.121	0.105
HGS-Sched	**0.021**	1	1	1	1	1	1	**< 0.0001**
IGA	**0.052**	**0.041**	**0.039**	**0.031**	0.121	**0.032**	0.059	0.101
GA+TS	1	**< 0.0001**	**0.027**	**0.031**	**0.023**	**0.0029**	0.198	1

Table 4.19 contains the *P-value* parameters, which are interpreted as the probabilities of the errors of Type 1 [101]. These results give the adequate information about possible acceptance or the rejection of the 'null' hypothesis. The difference in the results is not statistically significant if the *P-value* is not greater than 0.05 (*P-value* is 1 for the base (best) results).

In the static instances the results achieved by the single-population algorithm differ significantly from the results generated by the other meta-heuristics. The efficiency of *HGS-Sched* and *TS+GA* is similar in all but large-scale grids in the case of flowtime minimization. IGA algorithm is more effective than single-population scheduler, but it is not as good as hierarchical and hybrid meta-heuristics. In the dynamic scenario the highest effectiveness of $HGS-Sched$ is demonstrated in all but very large grid in the case of flowtime optimization. The differences in the achieved $Makespan$ results are small, while in the case of flowtime the *HGS-Sched* is the best adapted to the system dynamics.

4.5 Summary

This chapter presents the comprehensive experimental analysis of various implementations of *HGS-Sched* grid scheduler versus Island GA, single-population and hybrid GA meta-heuristics. Those techniques were used for exploration of the bi-objective fitness landscape for independent grid scheduling. Hybrid methodology is based on the 2-level search of *GA* and TS algorithms. *GA* plays the role of the control strategy, while TS in subordinated module is responsible for the accurate exploration of the neighborhoods of suboptimal schedulers detected by *GA*.

The results of experiments show the high effectiveness of $GA+TS$ hybrid meta-heuristic in static scheduling. Indeed, TS component is efficient in the fast reduction of the completion times of machines. It can also easily detect the near-optimal local solutions distributed in the small narrow regions of the search space. On the other hand, the major drawback of this method may be the low resistance to the system dynamics. In such cases *HGS-Sched* outperforms the other approaches. Hierarchical genetic scheduler is the most efficient in all considered scenarios of dynamic grids, which makes it a good candidate solution for scheduling in the realistic grids. The general *HGS-Sched* framework presented in Chapter 3 was successfully applied for the implementation of single- and multi-population genetic schedulers. Different models of the schedulers can be created just by setting a proper configuration of the key parameters in the main hierarchical framework.

Similarly to the classical combinatorial optimization, even simple theoretical analysis of the scheduling landscapes in computational grids may be a strong support to the design of the scalable grid schedulers that can be easily adapted to actual system states. This research area is still very superficially explored and each progress in the landscape characteristics may be the hottest research issue in future generation grid and cloud computing.

Part III
Security-Driven Scheduling Model for Computational Grid Using Multi-Level Genetic Metaheuristics

Chapter 5
Security-Aware Independent Batch Scheduling in Computational Grids

Abstract. This chapter presents a model for independent batch scheduling in Computational Grid that enables the aggregation of security requirements as additional scheduling criteria. Artificial Neural Network (ANN) module is an important component of this model. It is designed for supporting the security-aware evolutionary single- and multi-population grid schedulers.Based on a preliminary analysis of the trust levels of resources and security demand parameters of tasks, the neural network monitors the scheduling and task execution processes and generates the tasks-machines mapping "suggestions" based on the information about resource failures and the resulting tasks and machines characteristics. This information is used by the schedulers for an effective minimization of the scheduling objective function and the improvement of the system throughput.

5.1 Introduction

While the maximization of the resource utilization and profits of the resource owners are the key objectives of the grid scheduling, they may conflict with grid users' security requirements and system reliability . A major hurdle in effective job outsourcing in grid is caused by network security threats. The grid resources may not be accessible if the grid cluster is under attack. The system infections may lead to machine crashes during the execution of tasks dispatched to that cluster. Therefore, it is desirable to have a prior knowledge about the security demands from grid jobs and the trust level assured by a resource provider at the grid cluster. An effective grid scheduler must be then security-driven and resilient in response to all scheduling and risky conditions. It means that to achieve the successful tasks executions according the specified users' requirements, the relation between the assurance of secure computing services by a grid site or by a cluster node (security) and the behavior of a resource node (trust) must be defined and analyzed.

The main problem addressed in this chapter is an improvement of the effectiveness of the single- and multi-population genetic-based grid schedulers in the low-cost resource allocations under security constraints. The security awareness of those

J. Kołodziej: Evolutionary Hierarchical Multi-Criteria Metaheuristics, SCI 419, pp. 81–111.
springerlink.com

schedulers is supported by an *Artificial Neural Network (ANN)* module integrated with the system. Based on a prior analysis of trust levels of the resources and security demand parameters of tasks, the neural network monitors the scheduling and task execution, and produces task-machine mapping "suggestions" (recommendations) by using the system information, such as resource failure rates and system input parameters. Thereafter, based on the ANN "suggestions", sub-optimal schedules are generated and used in the initialization procedures of genetic-based schedulers for optimizing the main scheduling objective functions such as makespan and flowtime.

Despite the generation of the sub-optimal solution to the specified scheduling problems, the ANN module is *not considered* in this work as additional scheduler. It works in a "background" of the main scheduling process and monitors the scheduling results. However, the schedules generated by ANN may be accepted as the optimal solutions if the employed schedulers cannot generate the better ones.

According to the notation introduced in Sec. 1.4.2 the independent batch security-aware scheduling in which the makespan and flowtime are optimized in a hierarchical mode can be specified as follows:

$$Rm\left[\{b, indep, (stat, dyn), hier\}\right]\left(C_{max}(sec)\left[C_{max}(ris)\right], F(sec)\left[F(risk)\right]\right)\right) \quad (5.1)$$

where:

- $C_{max}(sec)$ – stands for a makespan as the primary scheduling objective under security constraints;
- $F(sec)$ – stands for a flowtime as the second scheduling objective under security constraints;
- $C_{max}(risk)$ – stands for a makespan as the primary scheduling objective in the risky scheduling mode;
- $F(risk)$ – stands for a flowtime as the second scheduling objective in the risky scheduling mode.

The procedures of calculating the $C_{max}(sec)$, $F(sec)$, $C_{max}(risk)$ and $F(risk)$ values will be defined later on. The interpretations of the remaining parameters are the same as in Sec. 2.1 (Eq. (2.1)).

This chapter extends the model and results presented in [18] by the implementation and the comparative analysis of the effectiveness of multi-population and single-population GA-based grid schedulers and the integration of the ANN module with *Sim-G-Batch* grid simulator. In the ANN module the *Minimal Completion Time (MCT)* algorithm is used for the generation of sub-optimal schedules.

5.2 Related Work

There has been a number of studies over the last years in which the security procedures in grid scheduling are verified in risky environments, where the resource trust parameters must be analyzed. The security-aware scheduling process in grid environment [65], [152], [165] is more difficult for the management than conventional scheduling defined for supercomputers, real-time, and parallel computers [66], [91],

[100]. Unfortunately, well-known scheduling approaches for grid computing largely ignore this security factor, with only a handful of exceptions.

A simple classification of security-aware grid models for an immediate job execution mode is presented by Humphrey and Thompson in [65]. They define a job control system for accessing grid information services through authentication. However, they did not elaborate on how a scheduler should be designed to address the security concerns in collaborative computing over distributed cluster environment. An extensive survey of the research endeavors in this domain is presented in [34].

Hwang et al. [67] developed an interesting fault-tolerance mechanism in CGs with a failure detection service, that enables the detection of both task failures and user's secure requirements in a dynamic environment. Abawajy [1] developed a model, that faces the system dynamics by a replication of the users' jobs at multiple grid sites in order to improve reliability of grid resources, and successful job executions.

Due to their high scalability heuristic methods seem to be the effective tools in solving the large-scale grid scheduling problem with additional security and resource reliability criteria [137]). However, security and task abortion mechanism are usually applied as the external procedures separated from the core of the scheduling system. For example, security requirements can be specified in the grid system by using a simple trust model [8].

Some recent security-aware approaches in CG scheduling are based on the game-theoretical models . In [136] and [137] the authors define the risky and secure conditions in online scheduling in CGs caused by software vulnerability and distrusted security policy. They apply the game model introduced in [92] for simulating the resource owners selfish behavior. The results presented in [137] are extended by Wu et al. in [155]. The authors consider the heterogeneity of fault-tolerance mechanism in a security-assured grid job scheduling and define four types of GA-based online schedulers for the simulation of fault-tolerance mechanisms.

In the aforementioned models the final decisions on the secure allocation of task to resources are made by the CG users who do not cooperate with each other. The costs of the risk-resilient tasks executions are interpreted as the users' cost functions, which are specified as the scheduling objectives and are minimized during the game. The main drawback of the online scheduling approaches may be the high computational complexity of the schedulers. In many cases the games are provided on the different grid levels and the design of an effective synchronization mechanism is a challenging task. A game-theoretical support to the users' decisions and actions will be discussed in Chapter 6.

Artificial Neural Networks (ANNs) are usually implemented as schedulers in grid computing. An illustrating example can be the grid scheduler based on the Fuzzy Neural Networks presented in [166]. The authors used the fuzzy logic module for monitoring the status of machine loads in grid system. The parameters of fuzzy membership functions in this model are tuned by using the ANN trained by back-propagation algorithm.

In [131] the ANN mechanism is used for supporting the users' decisions. The authors defined a decision model which is composed of three main components: (a) online module for the prediction of the users' actions; (b) off-line module for the analysis of statistical data acquired during user's work; and (c) 'users activity' module defined for the detection of trends and changes in users' activities. The users' decisions mechanisms are supported by the feed-forward neural networks trained by the back-propagation method. The authors additionally proposed the offline model, where another neural network is applied for the detection of normal/abnormal users activities, by analyzing the statistical data accumulated during the users' actions.

The above mentioned model is a promising solution for simulation and simple analysis of the users' decisions. This model can be considered as an alternative to game-based methodology in online scheduling. However the high complexity of this model can be a main drawback for its successful application in real-life grid scenarios.

5.3 Security as Scheduling Criterion in Computational Grids

A general security-aware grid model is based on the hierarchical multi-level architecture presented in Chapter 1 (see Sec. 1.2.2). However, the role of the meta-scheduler is different when security is considered as additional criterion in the scheduling process. The meta-scheduler must analyze the security requirements for the execution of tasks and requests of the CG users for trustful resources available within the system. The system brokers analyze "reputation" indexes of the machines received from the resource managers and send proposals to the scheduler. Moreover, the brokers also control the resource allocation and communication between CG users and resource owners.

Fig. 5.1 depicts the 3-level architecture of the security-aware grid cluster.

The trust level and security demand parameters are generated by aggregation of several scheduling and system attributes. Those parameters depend heavily on the security policy, accumulated resource or grid cluster "reputation", self-defense capability, attack history, special users' requirements, and peer authentication. Fig. 5.2 presents the major behavior and intrinsic security attributes needed for the specification of trust levels of the grid clusters and security demand of the grid applications (see also [137]).

Song et al. in [136] have developed a fuzzy-logic trust model, in which the aforementioned attributes are aggregated into single scalar parameters. The task security demand in this model is supplied by the user's programs as request for authentication, data encryption, access control, etc. The trust level parameters of the resource clusters are aggregated through a **two-level** hierarchic fuzzy-logic based trust procedure in the following way:

- At the lower *intra-site* level there are applied two fuzzy inference systems for the evaluation of the self-defence capabilities and trust indexes of the resources; each grid cluster reports its assessed self-defense capability to all other clusters;
- At the higher *inter-site* level there are collected the inputs from all resource clusters and the trust level vector is defined through another fuzzy inference process (see [136]).

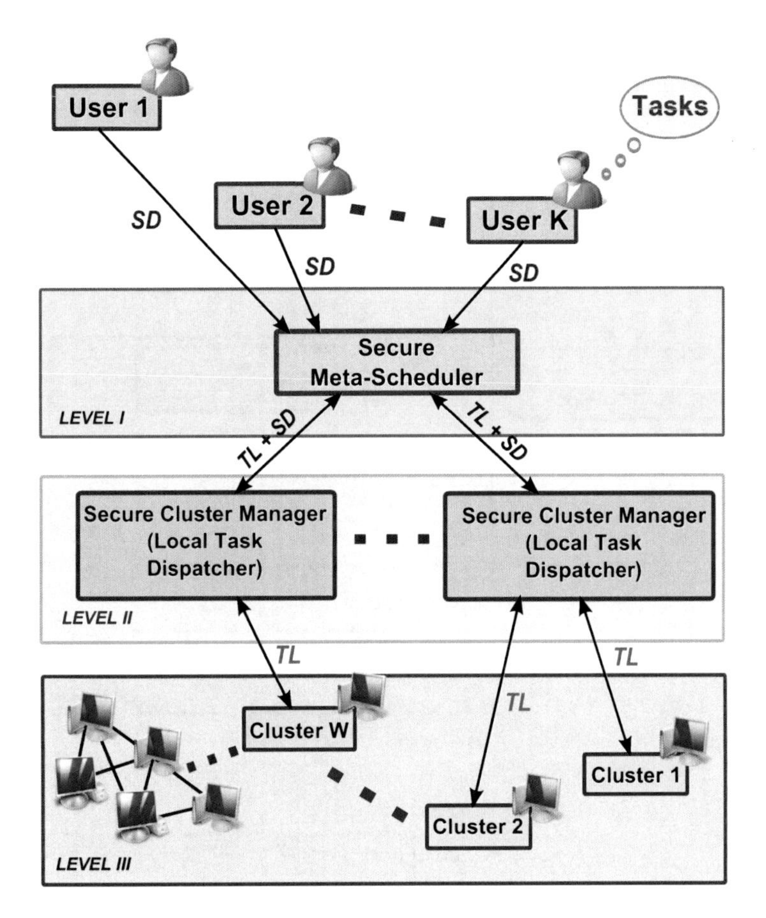

Fig. 5.1 The model of secure grid cluster

This fuzzy trust model was used in this work for the specification of new characteristics of tasks and resources in the grid system, namely security demand and trust level vectors. The *security demand vector* , denoted by $SD = [sd_1, \dots, sd_n]$ sd_j, is defined as a vector of the security demand parameters sd_j, $(j \in N)$, for all tasks in the batch. The *trust level vector* , denoted by $TL = [tl_1, \dots, tl_m]$, is defined as a vector of trust level parameters tl_i for all resources in the system. The trust level parameters specify how much a grid user can trust the resource manager. The manager maintains machine i status and monitors the execution of the tasks assigned to this machine. The values of the sd_j and tl_i parameters are real fractions within the range [0,1] with 0 representing the lowest and 1 the highest security requirements for a task execution and the most risky and fully trusted machine, respectively. A task can be successfully completed at a resource when a *security assurance condition* is satisfied. That is to say that $sd_j \leq tl_i$ for a given (j,i) task-machine pair.

Let us denote Pr_f to be a *Machine Failure Probability* matrix , the elements of which, are interpreted as the probabilities of failures of the machines during the tasks

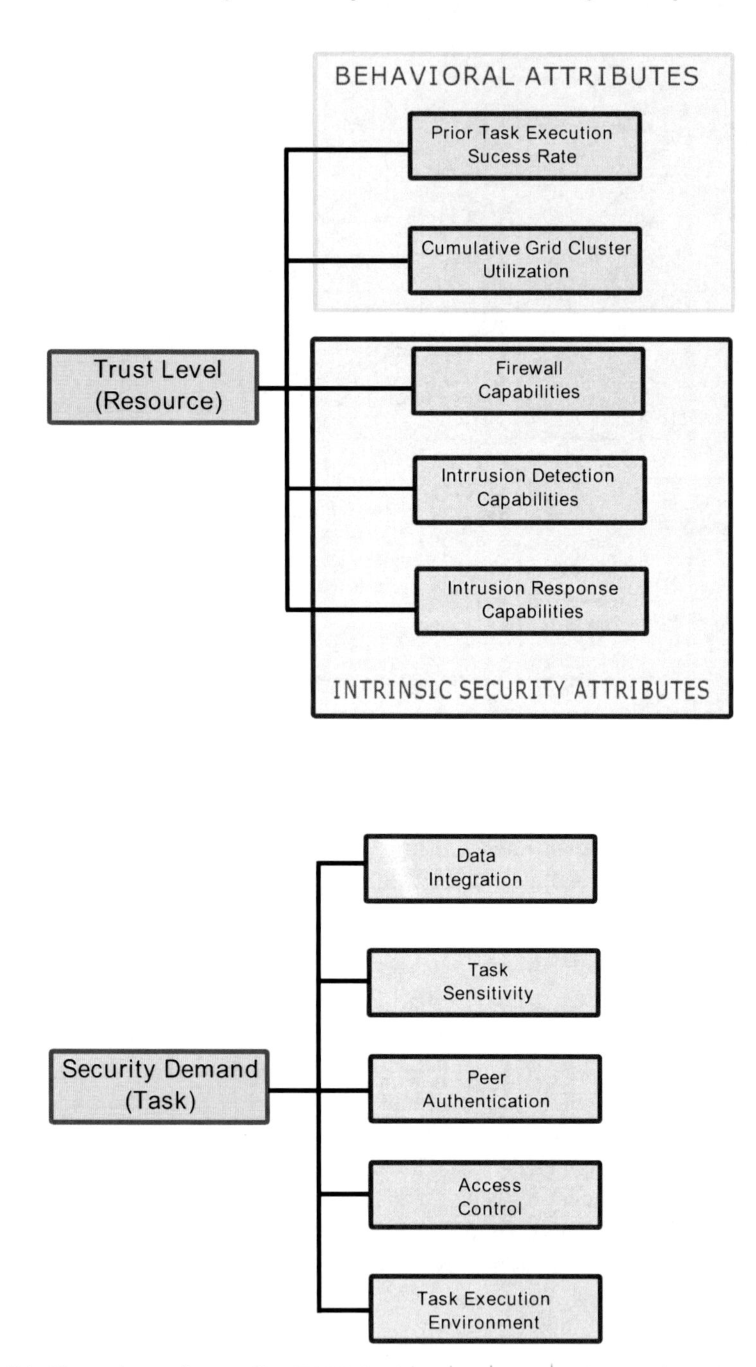

Fig. 5.2 The major attributes affecting the trust level and security demand in grid systems

executions due to the high security restrictions. These probabilities are denoted by $P_f[j]i]$ and are calculated by using the negative exponential distribution function, that is to say:

$$Pr_f[j][i] = \begin{cases} 0 & , sd_j \leq tl_i \\ 1 - e^{-\alpha(sd_j - tl_i)} & , sd_j > tl_i \end{cases} \quad (5.2)$$

where α is interpreted as a failure coefficient and is a global parameter of the model.

The process of matching sd_j with tl_i is similar to that of a real-life scenario where users of some portals, such as Yahoo!, are required to specify the security level of the login session.

5.3.1 Scheduling Scenarios and Objectives

The grid cluster or the grid resource may be not accessible to the global meta-scheduler when being infected with intrusions or by malicious attacks. The scheduler has two options of initializing his work: (a) to analyze the *Machine Failure Probability* matrix in order to minimize the failure probabilities for task-machine pairs; or (b) to perform an "ordinary" scheduling without any preliminary analysis of the security conditions, abort the task scheduling in the case of machine failure, and reschedule this task at another resource. The scheduler's strategies give rise to two modes of processing (and modelling in particular) the grid schedules, namely *secure* and *risky* modes.

Secure Mode

In this scenario all of the security and resource reliability conditions are verified for all task-machine pairs. The main goal of the meta-scheduler is to design an optimal schedule for which, beyond the makespan and flowtime, the probabilities of failures of the machines during the tasks executions will be minimal. It is assumed that additional "cost" of the verification of security assurance condition for a given task-machine pair: (a) may delay the predicted execution time of the task on the machine, and (b) is proportional to the probability of failure of the machine during the task execution. This "cost" is defined as the product of $Pr_f[j][i]$ and $ETC[j][i]$. In this case the completion time of the machine i is denoted by $completion^s[i]$[1] and can be calculated as follows:

$$completion^s[i] = ready_i + \sum_{\{j \in Tasks(i)\}} (1 + Pr_f[j][i]) ETC[j][i]) \quad (5.3)$$

where $Tasks(i)$ denotes the set of tasks assigned to the machine i in a given batch.

In this mode the main scheduling objectives, namely makespan and flowtime, can be expressed as follows:

[1] The general concept of the completion time of machine was explained in Sec. 2.2.2 in Chapter 2.

$$C_{max}(sec) = \max_{i \in M} completion^s[i]. \tag{5.4}$$

$$F(sec) = \sum_{i \in M} F^s[i] \tag{5.5}$$

where

$$F^s[i] = ready_i + \sum_{j \in Sorted[i]} (1 + Pr_f[j][i]) ETC[j][i]) \tag{5.6}$$

and $Sorted[i]$ denotes the set tasks assigned to the machine i sorted in ascending order by the corresponding ETC values.

Risky Mode

In this scenario all secure and failing conditions are ignored. The scheduling process is realized as a two-step procedure. First, the scheduling is performed just by analyzing the *ETC* matrix. If failures of machines are observed, then the unfinished tasks are temporarily moved into the backlog set. This set is defined as a 'batch supplement' and the tasks form this set are re-scheduled in the way as in the secure mode. The total completion time of machine $i (i \in M)$ in this case can be defined as follows:

$$completion^r[i] = completion[i] + completion^s_{res}[i] \tag{5.7}$$

where $completion[i]$ is calculated by using the Eq. (2.13)(see Chapter 2, Sec. 2.2.2), for tasks primarily assigned to the machine i , and $completion^s_{res}[i]$ is the completion time of machine i calculated by using the Eq.(5.3) for rescheduled tasks, i.e. the tasks re-assigned to the machine i from the other resources.

The formulas for makespan and flowtime in this mode are defined in the following way:

$$C_{max}(risk) = \max_{i \in M} completion^r[i]. \tag{5.8}$$

$$F(risk) = \sum_{i \in M} F^r[i] \tag{5.9}$$

where

$$F^r[i] = ready_i + \sum_{j \in Sorted[i]} ETC[j][i] + \sum_{j \in Sorted_{res}[i]} (1 + Pr_f[j][i]) ETC[j][i]) \tag{5.10}$$

and and $Sorted_{res}[i]$ denotes the set of *rescheduled* tasks assigned to the machine i sorted in ascending order by the corresponding ETC values

Assuming the hierarchical optimization mode (see Eq. (5.1), parameter *hier*) with the makespan as the primarily scheduling criterion, the flowtime should be minimized in both secure and risky scenarios subject to the the following constraints:

- in the secure mode

$$F^s[i] \leq C_{max}(sec) \; \forall i \in M; \tag{5.11}$$

- in the risky mode

$$F^r[i] \leq C_{max}(risk) \; \forall i \in M. \tag{5.12}$$

Although the probabilities of machines' failures are expected to be higher in the risky than in the secure mode, there is certainly no guarantee of the successful execution of all tasks in the security scenario. It can be observed that if the *security assurance condition* is satisfied for each task-machine pair (i.e. $sd_j \leq tl_i$ for $i \in M, j \in N$), the completion times of machines in both *secure* and *risky* modes are identical with the completion times defined for standard independent scheduling problem (see Chapter 2, Eq.(2.13)), where it is assumed that each task *must* be successfully executed on each machine and no security requirements are analyzed[2].

5.4 Artificial Neural Network Module

The implementation of Artificial Neural Network (ANN) module requires preliminary classification of tasks and machines available in the system. This classification is based on the values of the workload (WL), computing capacity (CC), trust level (TL) and security demand (SD) vectors. Machines are categorized into the R_r types according to their *processing power* features, namely *slowest, slower,$\cdots$,medium,$\cdots$ fastest* classes; and into R_s types according their *trust level* features, namely *secure, less_secure,$\cdots$,medium,$\cdots$ fully_risky* classes. This initial classification leads to the overall categorization of the resources into $R = R_r \cdot R_s$ classes, namely *slowest-secure,$\cdots$,medium-secure, fastest-fully_risky* types, in order to characterize the grid machine under the processing power and trust criteria.

Similar classification can be provided for the submitted tasks under workload and security demand features. The tasks are categorized into $T = T_w \cdot T_{sd}$ types, where T_w is number of *workload* classes and T_{sd} is number of *security demand* classes. R machine classes and T task classes generate $R+T$ possible inputs for neural network.

Formally, the ANN input data can be expressed by the following pair of vectors: $\{TASKS\_MX; MACHINES\_MX\}$, where:

- $TASKS\_MX[\hat{t}] = T_t$ for tasks classification, where $\hat{t}$ denotes a task class, ($t = 1,\ldots,T$), and T_t denotes the fraction of tasks in the class $\hat{t}$. That is to say:

$$T_t = \frac{\widehat{t_t}}{n}, \tag{5.13}$$

[2] The component $completion^s_{res}[i]$ in Eq. (5.7) is not calculated while all tasks are successfully executed on the grid machines.

where $\widehat{t_t}$ is the number of tasks in the class $\widehat{t}$ and

$$\sum_{t=1}^{T} T_t = 1 \tag{5.14}$$

- $MACHINES\_MX[\widehat{r}] = R_r$ for resources classification, where $\widehat{r}$ denotes a machine class ($\widehat{r} = 1,\ldots,R$), and proportion of machines in the class $\widehat{r}$ is denoted by R_r. That is to say:

$$R_r = \frac{\widehat{r_r}}{m}, \tag{5.15}$$

where $\widehat{r_r}$ is the number of machines in the class $\widehat{r}$ and

$$\sum_{r=1}^{R} R_r = 1 \tag{5.16}$$

ANN module monitors the machine failures and the successful execution of tasks on machines. The information about the failures of the grid resources is needed for the classification of the results generated by the neural network. For each class $\widehat{r}$ of machines, there is selected a unique *major class* $t_{maj}(\widehat{r})$ of successfully executed tasks. This class is specified based on the number of completed tasks on a given machine without any failures and re-scheduling procedures. The results generated by the neural network are defined as an *output matrix* OUT_MX of the size $T \cdot R$ with just R non-zero (positive) elements (one major class of tasks for each host is indicated in such a way), where:

$$OUT\_MX[\widehat{r}][t_{maj}(\widehat{r})] = r_{(t_{maj}(\widehat{r}))} \tag{5.17}$$

and $r_{(t_{maj}(\widehat{r}))}$ is the proportion of the tasks from the major class $t_{maj}(\widehat{r})$ submitted to the machines of the class $\widehat{r}$. The main concept of the network is presented in Fig. 5.3.

The network is trained by the *back propagation* algorithm [61]. The results generated by the ANN module are used for the specification of the grid schedules, which may be accepted as the partial (or optimal) solutions for the problem, or may be passed on to the heuristic or meta-heuristic schedulers as the initial solutions. The procedure of the generation of the schedules based on the ANN "suggestions" can be defined as follows. First, a 'major class' $t_{maj}(\widehat{r})$ membership is verified for all tasks in the batch. For each task from the 'major class' the class of the fastest and most trustful machines is selected. Thereafter, this task is assigned to the machine from the selected class, with the minimal completion time. The tasks from the other classes than the 'major' one are assigned to the machines with the shortest completion times without analyzing the network results. The *Minimum Completion Time (MCT)* method is used for all those assignments. The general framework of MCT procedure is presented in Alg. 2.

Both input and output matrices defined for ANN are totally independent of numbers of hosts and tasks in the system. Therefore, the ANN module can be trained even on a small batch of task and small cluster of machines and then the generated results may be used in more complex scenarios.

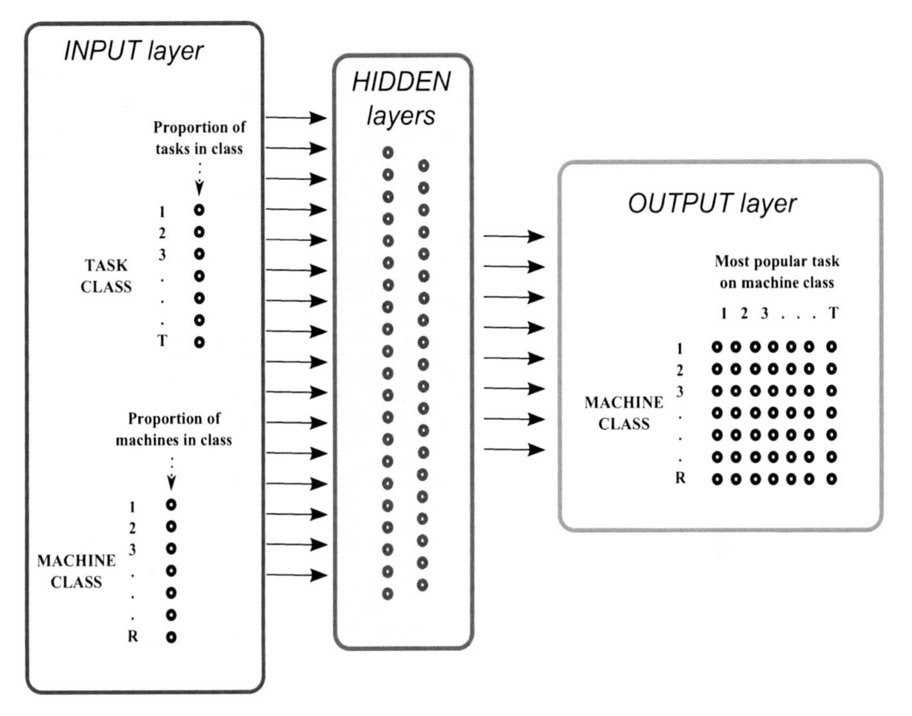

Fig. 5.3 Neural network architecture

Algorithm 2. *MCT* algorithm template

1: Calculate the *ready_times* of the machines ;
2: **for all** Tasks in a given batch **do**
3: Calculate the completion times of the machines for the tasks;
4: Find the machine that gives minimum completion time, m_{best};
5: Assign task to m_{best} machine;
6: Update the machine completion time;
7: **end for**
8: Return the resulting schedule

5.5 Empirical Evaluation of the Genetic Metaheuristics for Security-Aware Scheduling

The specification of just one major class of tasks for the network results may be of course the main reason of relatively low efficiency of the ANN module in some realistic approaches. The intelligent modification of this neural network model into the low-cost (in the sense of the execution time) multi-class version remains still challenging research issue. However, the monitoring of the system, the provisioning of the resources and scheduling according to various security requirements, should be even partially automated. Otherwise, the information about the reliability of the resources, all negotiations of the particular security conditions, or simply the information about the failures of the nodes, must be proceed by the system administrators and grid users. It usually delays the realization of the schedules because of the sheer scale of the system and different, and often conflicting, goals and interests of the users working at the various system levels. The ANN model presented in this chapter may serve as a prototype solution for both automatic monitoring and detection of the system failures and intelligent supporting of the users decisions, with a tangible benefit of reducing an overall computing overhead in high-performance computing systems (not just grids).

The aim of the empirical study presented in this section is the verification of the efficiency of the neural network support for single- and multi-population genetic-based schedulers in the reduction of the number of failures of the resources caused by too restrictive security conditions, and in the minimization of two conventional scheduling objectives, namely makespan and flowtime. The experiments were conducted in two main steps. First, six variants of single-population GA-based schedulers with different crossover, mutation and replacement operators were evaluated in both scheduling modes. The goal of this analysis is to compose an optimal combination of genetic operators for multi-population hierarchical, island and hybrid genetic schedulers. Thereafter, four genetic meta-heuristics were evaluated in risky and secure modes, namely the best single-population GA in the first part of the analysis, and *HGS-Sched*, *Island GA* and hybridization of GA with Tabu Search (*GA+TS*).

5.5.1 Security Aware Sim-G-Batch Grid Simulator

The ANN module was integrated with the *Sim-G-Batch* simulator for modelling and monitoring the grid system behavior under the specified security conditions. The module works in the "background" of the main system and supports the resolution methods implemented in *Scheduler* class. The sub-optimal schedules generated based on the neural networks "suggestions", are passed on to the initial populations of the genetic schedulers. The main concept of the security-aware version of *Sim-G-Batch* simulator with the ANN module is presented in Fig. 5.4.

In the case of security scheduling, the list of typical input parameters for *Sim-G-Batch* (see Chapter 2, Sec. 2.3.1) is extended by the trust level vector TL and the security demand vector SD. Table 5.1 presents the values of key parameters in four grid size scenarios, namely *Small*, *Medium*, *Large* and *Very Large* in static and dynamic modes. Most of those parameters (excluding the numbers of tasks and machines) were tuned in empirical analysis presented in [136], [137], [92] and the recent publications of the author of this book [18], [87], [89].

For activating the ANN module, the tasks and machines are divided into 18 classes: 9 categories for processing power and trust level criteria (machines), and 9 categories for workload and security demand criteria (tasks). The ANN is a feed-forward network with two hidden layers, the weight coefficients are in the range of $[-0.2;0.2]$ and the learning rate is 0.01. The training set for ANN contains the

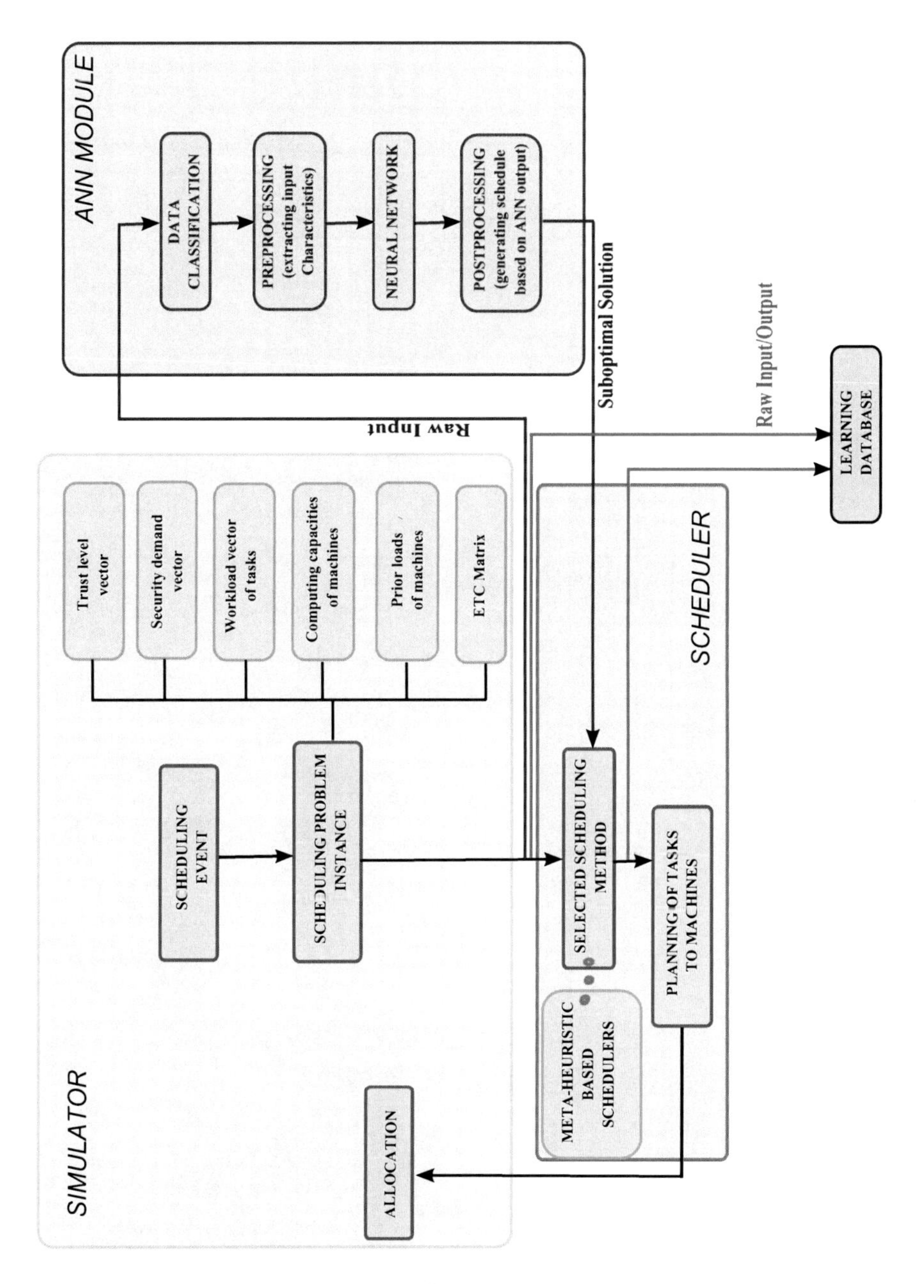

Fig. 5.4 General Flowchart of the secure *Sim-G-Batch* Simulator Supported by Neural Network Linked to Scheduling

Table 5.1 Values of key parameters of the secure *Sim-G-Batch* in static and dynamic modes

	Small	Medium	Large	Very Large
		Static case		
Nb. of hosts	32	64	128	256
Resource cap. (in MHz CPU)		$N(5000, 875)$		
Total nb. of tasks	512	1024	2048	4096
Workload of tasks		$N(250000000, 43750000)$		
Security demands sd_j		$U[0.6; 0.9]$		
Trust levels tl_i		$U[0.3; 1]$		
Failure coefficient α		3		
		Dynamic case		
Init. hosts	32	64	128	256
Max. hosts	37	70	135	264
Min. hosts	27	58	121	248
Resource cap. (in MHz CPU)		$N(5000, 875)$		
Add host	$N(625000, 93750)$	$N(562500, 84375)$	$N(500000, 75000)$	$N(437500, 65625)$
Delete host		$N(625000, 93750)$		
Init. tasks	384	768	1536	3072
Total tasks	512	1024	2048	4096
Inter arrival	$E(7812.5)$	$E(3906.25)$	$E(1953.125)$	$E(976.5625)$
Workload		$N(250000000, 43750000)$		
Security demands sd_j		$U[0.6; 0.9]$		
Trust levels tl_i		$U[0.3; 1]$		
Failure coefficient α		3		

characteristics of the tasks and machines and the task-machine matching results collected after the 500 runs of the simulator with inactive Neural Network module.

5.5.2 Performance Measures

The performances of all schedulers in experiments were evaluated under the following three metrics:

- *Makespan* – a primarily scheduling criterion, which is expressed in Eq. (5.8) in risky scenario, and in Eq. (5.4) in the secure mode,
- *Mean_Flowtime* – flowtime scheduling criterion, which is defined in Eq. (5.8) for the risky mode, and in Eq. (5.4) for the secure mode; and
- *FailureRate* $Fail_r$ parameter defined as follows:

$$Fail_r = \frac{n_{failed}}{n} \cdot 100\% \tag{5.18}$$

where n_{failed} is the number of unfinished tasks, which must be rescheduled[3].

Both *Makespan* and *Mean_Flowtime* measures are expressed in arbitrary time units specified for the scheduling.

5.5.3 Tuning the Genetic Engine for Multi-Population Batch Schedulers

In the first part of empirical study six single-population risk-resilient GA schedulers have been compared in order to define an effective genetic engine for multi-population meta-heuristics. The configuration of the genetic parameters for those six schedulers are presented in Table 5.2.

Table 5.2 Configuration of six single-population GA-based grid schedulers

Scheduler	Replacement method	Scheduling scenario
GA-SS-R	Steady State	Risky
GA-SS-S	Steady State	Secure
GA-SS-ANN	Steady State	Secure supported by ANN
GA-ST-R	Struggle	Risky
GA-ST-S	Struggle	Secure
GA-ST-ANN	Struggle	Secure supported by ANN

The aforementioned methodologies differ in the implementation of the replacement mechanism. The *Steady State* replacement method is used in *GA-SS-xxx*

[3] According the notation introduced in Chapter 2, n stands for the number of tasks in a given batch.

algorithms and *Struggle* procedure – in *GA-ST-xxx*. The ANN module is active just in *GA-SS-ANN* and *GA-ST-ANN* algorithms for generating a part of an initial population. All of the remaining procedures in these algorithms are identical with the schedulers working in the secure scenario. Based on the results of the tuning process of genetic-based meta-heuristics in conventional grid scheduling presented in Chapter 4, Sec. 4.4.1, the remaining genetic operations in the schedulers are configured as follows: (i) *Linear Ranking* as selection scheme, (ii) *Cycle Crossover (CX)* operator and (iii) *Move* mutation method [107].

The generic frameworks of all considered schedulers are the same as in Alg. 1 defined in Chapter 3. The key parameters of HGS-Sched model for generating all types of genetic-based schedulers are presented in Table 5.3.

Table 5.3 GA Steady State and GA Struggle settings in static and dynamic cases

Parameter	Value
degree of branches (t)	0
period_of_metaepoch (α)	$(5*n)/10$
nb_of_metaepochs	10
population size (pop_size)	60
intermediate pop.	48
cross probab.	0.9
mutation probab.	0.15
$max\_time\_to\_spend$	200 sec. $(static)$ 400 sec. $(dynamic)$

The *HGS-Sched* has in this case just one core branch. Each experiment was repeated 30 times under the same configuration of operators and parameters.

5.5.3.1 Results

The results of the minimization of the *Makespan* in both *risky* and *secure* modes and all grid scenarios, are presented as the box-plots in Fig. 5.5–5.8.

The best results in the *Makespan* optimization have been achieved by both *GA-XX-ANN* schedulers. The efficiency of the ANN support can be observed especially

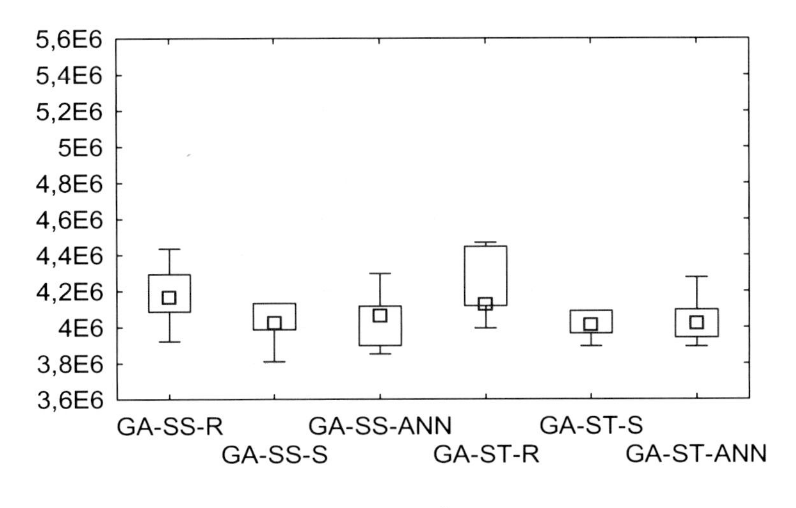

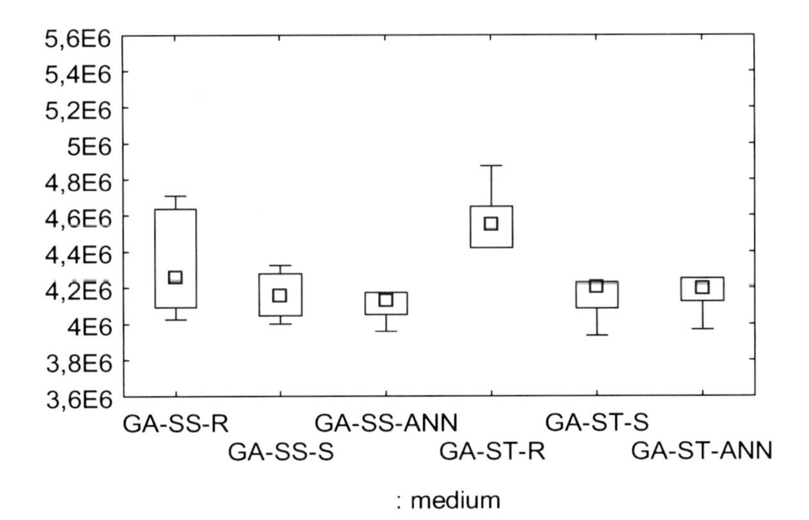

Fig. 5.5 The box-plot of the results for *Makespan* in static scheduling: Small and Medium grids

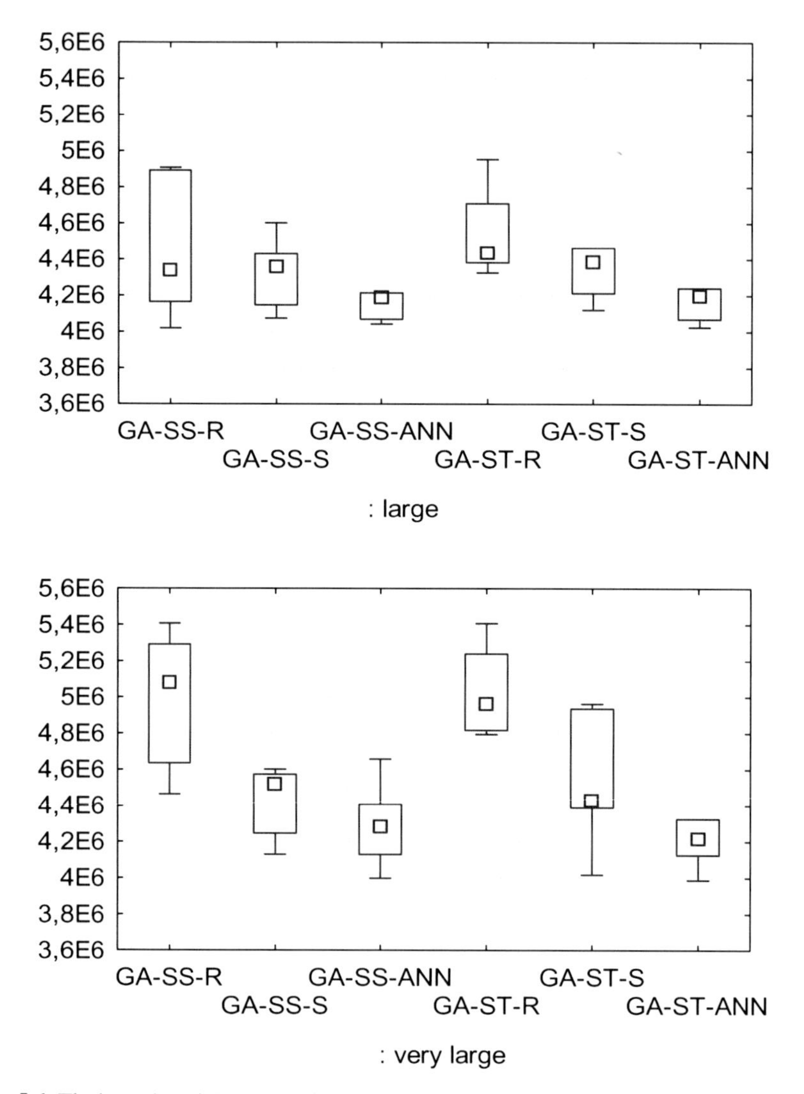

Fig. 5.6 The box-plot of the results for *Makespan* in static scheduling: Large and Very Large grids

in the 'Large' and 'Very Large' grid scenarios. The worst in the *Makespan* reduction were the schedulers working in the risky mode. While in the 'Small' grid case the differences in the averaged *Makespan* values are not so big, in all other scenarios *GA-SS-R* and *GA-ST-R* meta-heuristics significantly lag behind the secure schedulers. It can be also observed that in all instances the distribution of the results are asymmetric and the medians are very close to the first or the third quantiles.

The box-plots of the *Mean_Folwtime* results are presented in Fig. 5.9–5.12.

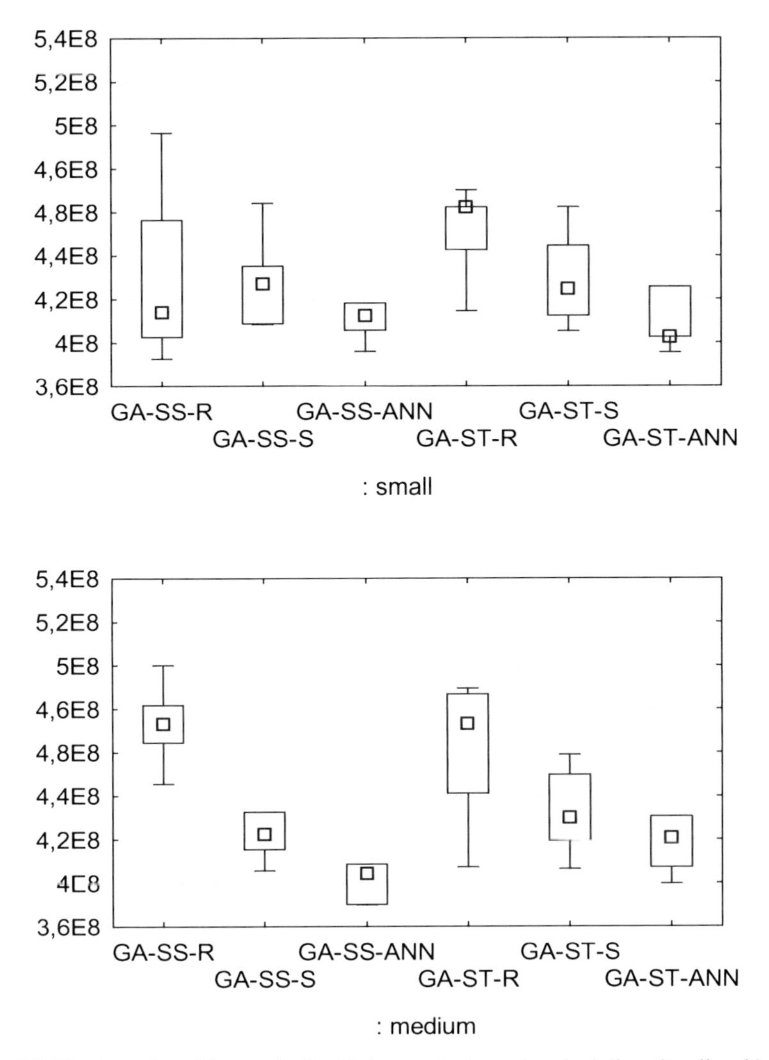

Fig. 5.7 The box-plot of the results for *Makespan* in dynamic scheduling: Small and Medium grids

In the case of the minimization of the flowtime, both *GA-XX-ANN* meta-heuristics outperform again the rest of the methods in both static and dynamic modes, however the differences in the results are not so significant as it was in the *Makespan* case. Additionally, it can be noted that as the instance size is doubled, the *Mean_Flowtime* values increase considerably for all applied schedulers, while the *Makespan* is almost at the same level. Another observation is that all schedulers are rather 'symmetric' in the sense of the distribution of the results and the differences between the first and the third quantiles are rather small. The best relative effectiveness of the ANN support may be observed in 'Very Large' grids in static and dynamic cases.

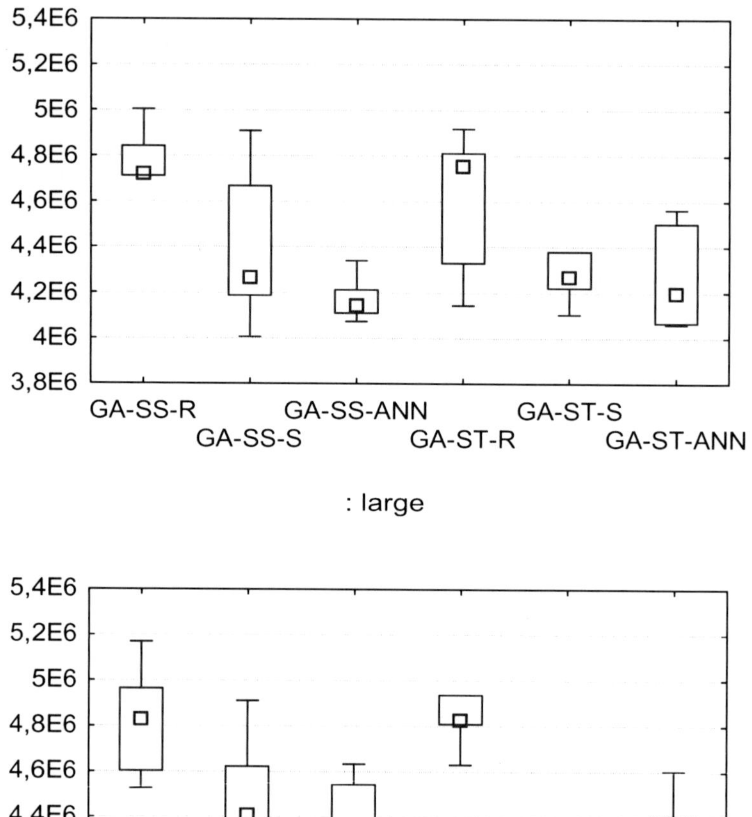

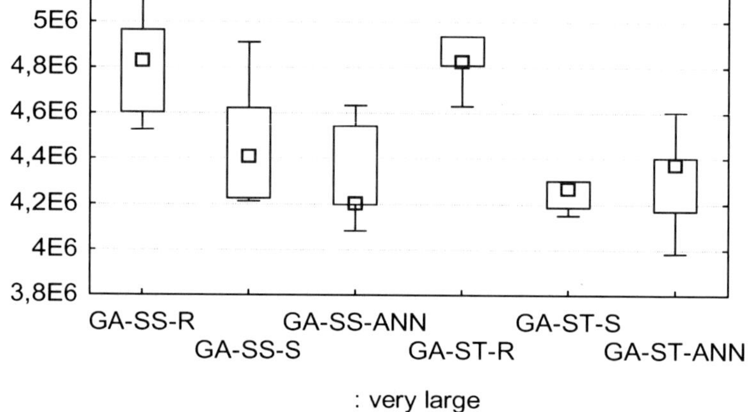

Fig. 5.8 The box-plot of the results for *Makespan* in dynamic scheduling: Large and Very Large grids

Similarly to other experiments presented in this book, the maximal number of generations in GA was defined as a stopping criterion for all schedulers. However, the solutions generated by ANN may not be improved by the GA meta-heuristics, and the search process can be stopped because of the stagnation in the improvement of the solutions' quality. Table 5.4 presents the averaged (in 30 runs of the simulator) minimal numbers of genetic epochs (generations) needed for generating the best solutions by all considered genetic schedulers. A relative effectiveness of each

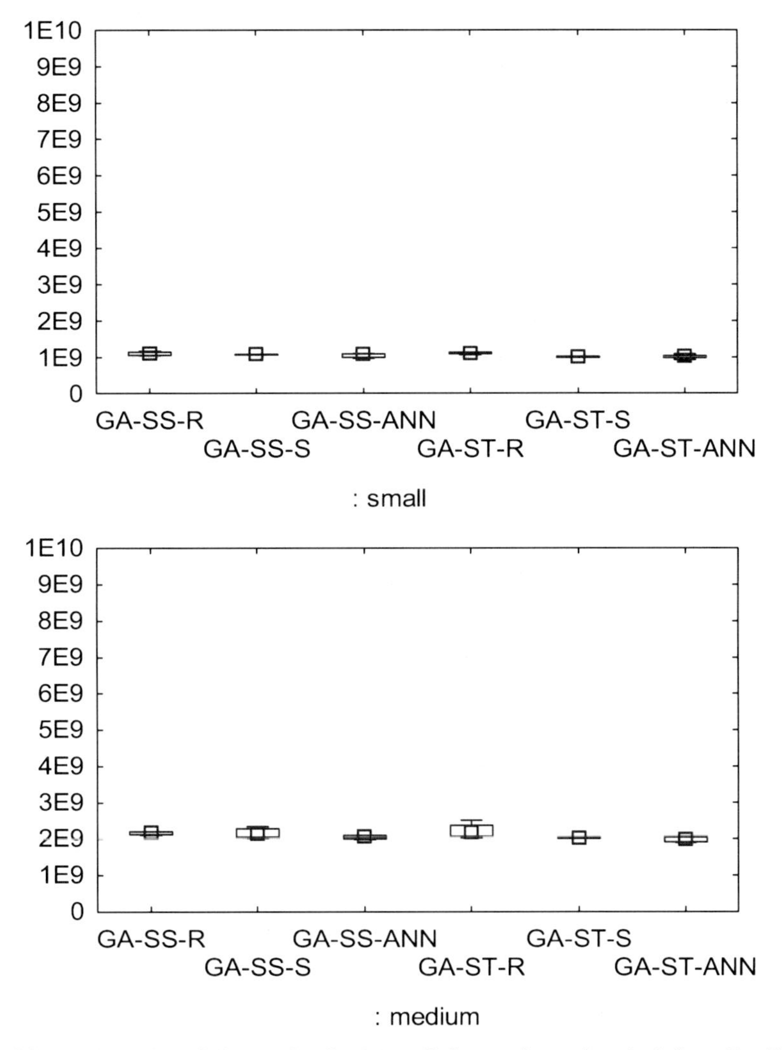

Fig. 5.9 The box-plot of the results for *Mean_Folwtime* in static scheduling: Small and Medium grids

scheduler is expressed as the ratio of the minimal number of genetic epochs necessary for finding the optimal solutions, and the stopping criterion, which is $5 \cdot n$, where n denotes the number of tasks in the system. These parameters are displayed in parentheses in Table 5.4.

It can be noted than ANN module in most of the instances reduced the time necessary for finding the best solutions approximately by 30–40 %, and successfully speeded up the search process in both secure and risky scenarios. The effectiveness of the ANN support is confirmed by the lowest failure rates achieved by the *GA-XX-ANN* schedulers. The results for all six schedulers are presented in Table 5.5.

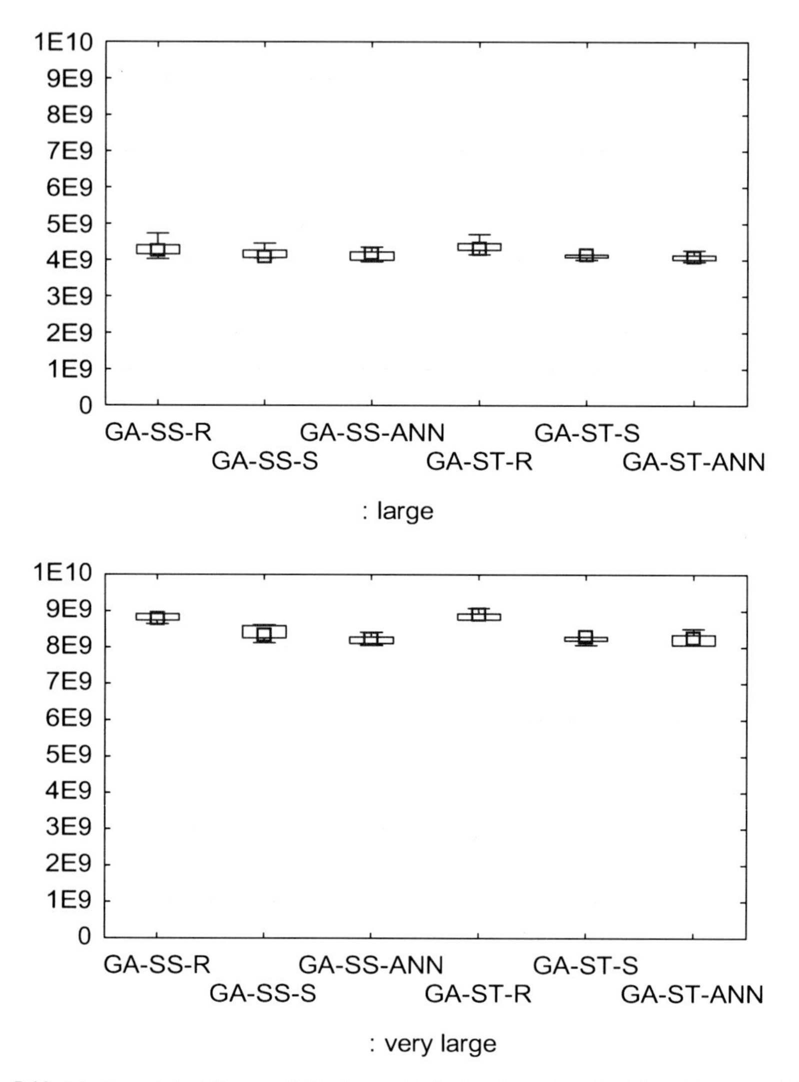

Fig. 5.10 The box-plot of the results for $Mean\_Folwtime$ in static scheduling: Large and Very Large grids

In all instances but one – the 'Small' grid in static scenario – the schedulers with the active ANN module outperform the other methods. The ANN support allow to reduce the machine failures by $1\% - -6\%$ compared to the 'conventional' schedulers.

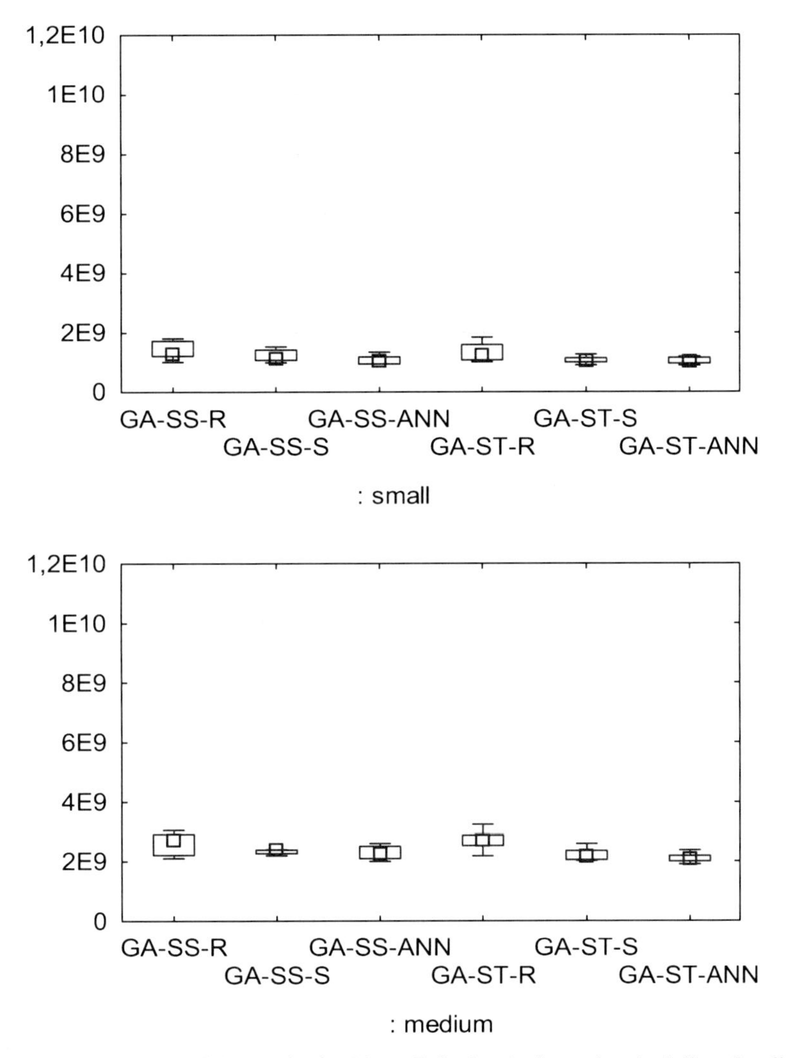

Fig. 5.11 The box-plot of the results for *Mean_Folwtime* in dynamic scheduling: Small and Medium grids

5.5.3.2 Summary

Based on the results of all experiments provided for single-population GA schedulers, the most effective in the optimization of both scheduling objective functions is *GA-SS-ANN*. This algorithm works in the secure mode with the ANN support and *steady state* replacement mechanism in the main genetic engine. This algorithm achieved the lowest failure rates in half of the instances, and most of them in the dynamic grid, which makes it the best candidate methodology for the secure

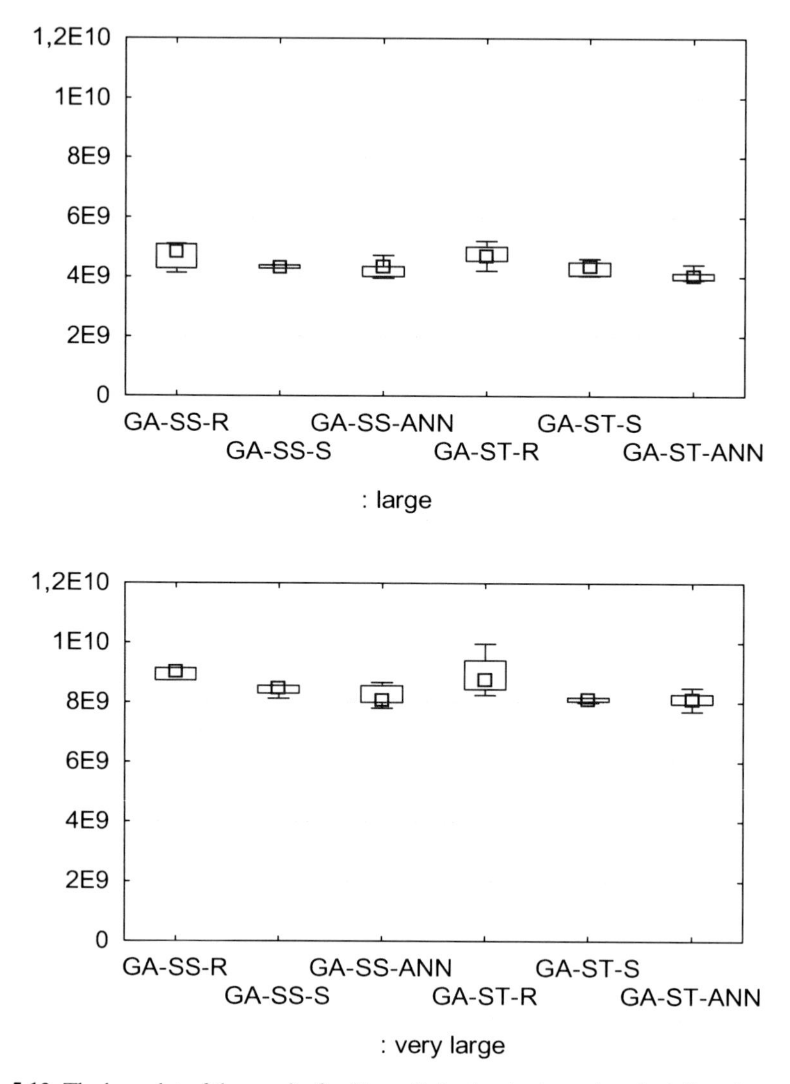

Fig. 5.12 The box-plot of the results for *Mean_Folwtime* in dynamic scheduling: Large and Very Large grids

scheduling in the realistic scenario. It is also the best in the minimization of the *Makespan* and flowtime in most of the instances of the scheduling problem.

It can be observed that generally it is more resilient for the grid schedulers to 'pay' a priori some additional scheduling 'cost' due to verification of the security conditions, than taking a risk on allocating the unreliable resources. As a result, the failure rates in the risky mode are much higher than in the secure case, especially in the dynamic grid where the frequency of the machine failures are 3–4 times greater than in the secure scenario. This is the main reason of lower effectiveness of

Table 5.4 The averaged minimal numbers of genetic epochs necessary for generating the best solutions by six considered GA-based schedulers

Strategy	**Small**	**Medium**	**Large**	**Very Large**
		Static Instances		
GA-SS-R	2302 (89.92%)	4722 (92.22%)	10008 (97.73%)	19226 (93.87%)
GA-SS-S	2031 (79.33%)	3620 (70.70%)	8345 (81.49%)	19740 (96.38%)
GA-SS-ANN	1722 (67.26%)	**2733** (53.37%)	**7992 (78.04%)**	17739 (86.61%)
GA-ST-R	1923 (75.11%)	4213 (82.28%)	10013 (97.78%)	20054 (97.91%)
GA-ST-S	1987 (77.61%)	4005 (78.22%)	8022 (78.33%)	18654 (91.08%)
GA-ST-ANN	**1592 (62.18%)**	3872 (75.62%)	8591 (83.89%)	**16940 (82.71%)**
		Dynamic Instances		
GA-SS-R	2090 (83.6%)	5099 (99.78%)	10100 (98.63%)	20145 (98.75%)
GA-SS-S	1831 (71.52%)	3925 (76.96%)	9036 (90.36%)	19002 (92.78%)
GA-SS-ANN	**1522 (60.60%)**	3021 (59.23%)	8010 (78.52%)	**17830** (87.06%)
GA-ST-R	2175 (85.21%)	4923 (96.15%)	10057 (98.59%)	19353 (94.86%)
GA-ST-S	1703 (68.52%)	**2954 (57.92%)**	8238 (80.70%)	17993 (88.63%)
GA-ST-ANN	1611 (61.44%)	3401 (66.68%)	**6035** (60.78%)	17910 (87.83%)

the schedulers in the optimization of the main grid objective functions in the risky mode. The ANN support in the security scheduling allow to reduce significantly the *Makespan* and *Mean_Folwtime* values and to keep the failure rates of the machines at the sufficiently low levels.

Table 5.5 Average values of failure rate $Fail_r$ parameter for six GA-based schedulers [$\pm s.d.$] ($s.d.$ = standard deviation)

Strategy	Small	Medium	Large	Very Large
		Static Instances		
GA-SS-R	4.832% [± 0.97]	7.201% [± 0.78]	11.824% [± 1.26]	31.721% [± 3.28]
GA-SS-S	4.008% [± 1.15]	4.135% [± 1.27]	10.698% [± 3.26]	11.635% [± 3.13]
GA-SS-ANN	3.993% [± 0.98]	**4.089%** [± 1.56]	8.436% [± 1.67]	**8.736%** [± 2.09]
GA-ST-R	4.697% [± 1.71]	17.516% [± 3.39]	14.013% [± 4.08]	35.643% [± 6.73]
GA-ST-S	**3.897%** [± 0.96]	5.540% [± 1.89]	10.945% [± 1.63]	10.402% [± 3.42]
GA-ST-ANN	3.967% [± 0.79]	6.430% [± 0.63]	**6.11%** [± 1.28]	9.196% [± 2.77]
		Dynamic Instances		
GA-SS-R	12.126% [± 1.80]	25.306% [± 2.79]	31.342% [± 3.44]	25.794% [± 2.48]
GA-SS-S	6.104% [± 1.69]	6.916% [± 2.40]	9.507% [± 1.84]	8.943% [± 2.07]
GA-SS-ANN	4.880% [± 0.98]	**6.097%** [± 1.62]	**7.456%** [± 1.32]	**7.026%** [± 2.11]
GA-ST-R	26.797% [± 5.25]	22.96% [± 4.19]	29.227% [± 4.95]	29.380% [± 5.30]
GA-ST-S	9.218% [± 2.84]	7.623% [± 2.02]	8.084 % [± 2.49]	9.744% [± 2.69]
GA-ST-ANN	**4.093%** [± 0.97]	6.991% [± 1.44]	7.681% [± 1.33]	7.894% [± 2.41]

5.5.4 *Evaluation of Multi-Population and Hybrid Genetic Metaheuristics*

The effectiveness of the multi-population meta-heuristics and hybrid genetic schedulers depend on the efficiency of their single-population genetic engines. $GA-SS-ANN$ algorithm, as the best single-population GA in the first part of the empirical analysis, was selected to serve as the main genetic mechanism in HGS-Sched, Island GA and $GA+TS$ hybrid algorithms applied to the secure grid scheduling. The following four meta-heuristics were considered in this part of analysis:

- *GA-SS-ANN* - with the settings defined in Table 5.3;
- *Sec-HGS-Sched* - with $GA-SS-ANN$ engine and various population sizes and mutation rates in the branches of different degrees;
- *Sec-IGA* - Island Genetic Algorithm with $GA-SS-ANN$ as the basic mechanism in all sub-populations;
- *Sec-(GA+TS)* - hybrid scheduler with $GA-SS-ANN$ as the control strategy and Tabu Search (TS).

The general characteristics of hierarchical, island and hybrid schedulers are presented in Sec. 4.4.2.1 in Chapter 4. The settings for all considered meta-heuristics are the same as the values of global parameters for *IGA*, *HGS-Sched* and *GA+TS* presented defined in Tables 4.16, 4.17 and 4.18. It means that *Sec-HGS-Sched* is composed of one branch of degree 0 and branches of degrees 1.

5.5.4.1 Results

The results of the comparative analysis of the minimization of *Makespan*, *Mean_Folwtime* and the failure rates in static and dynamic instances are presented in Tables 5.6, 5.7 and 5.8. The results were averaged over the 30 runs of the simulator for the same configuration of schedulers and all parameters.

The results show the high effectiveness of the hierarchic scheduler in the security-aware scheduling. *Sec-HGS-Sched* achieved the best results in 80 % of the instances for all scheduling metrics. It is the best in the reduction of the failing rates in 7 cases, which makes this model a solid base for the development of the real-life scheduling strategies in the security mode. The ANN module is a good candidate technology for an automatic verification of the security condition. *Sec-HGS-Sched* algorithm needs also the shortest time measured in the genetic epochs for the detection of the best schedules, which is illustrated in Table 5.9. This algorithm is the best in 7 instances, and the execution time for this method may be reduced in 25%–59% in the static case and in 21%–40% in the dynamic case.

Table 5.6 Average values of *Makespan* for single-population, multi-level and hybrid genetic schedulers [$\pm s.d.$], ($s.d.$ = standard deviation)

Strategy	Small	Medium	Large	Very Large
		Static Instances		
GA-SS-ANN	4208842.037 [± 210505.265]	4216980.163 [± 249225.887]	4309539.605 [± 263233.057]	4399950.825 [± 290453.201]
Sec-HGS-Sched	**3902040.474 [± 249630.764]**	**4051566.475 [± 319691.981]**	**4101943.296 [± 308590.795]**	**414387056.050 [± 2664631.423]**
Sec-IGA	4000936.859 [± 271909.245]	4208675.544 [±292686.570]	4245347.850 [± 328969.468]	4377434.150 [± 339217.338]
Sec-(GA+TS)	4070923.243 [± 282963.771]	4195886.584 [± 249817.482]	4278491.285 [± 262374.619]	4400502.382 [± 281474.189]
		Dynamic Instances		
GA-SS-ANN	4141538.885 [± 24798859.145]	4212439.475 [± 342459.080]	4232327.490 [± 333199.727]	4364692.950 [± 339043.674]
Sec-HGS-Sched	**3971502.411 [± 259973.626]**	**3991503.974 [± 321385.198]**	**4198873.263 [± 251572.072]**	**4227569.385 [± 281680.755]**
Sec-IGA	4064692.950 [± 339043.674]	4162170.950 [± 302537.087]	4202452.157 [± 277217.723]	4315327.490 [± 344120.912]
Sec-(GA+TS)	4039043.535 [± 281858.929]	4068783.648 [± 253899.771]	4230791.746 [± 290132.215]	4263913.826 [± 294304.036]

Table 5.7 Average values of *Mean_Folwtime* for single-population, multi-level and hybrid genetic [$\pm s.d.$], ($s.d.$ = standard deviation)

Strategy	Small	Medium	Large	Very Large
		Static Instances		
GA-SS-ANN	1098725220.445 [± 148984029.042]	2261958805.835 [± 196213971.853]	4395864089.470 [± 103819795.484]	8705728350.062 [± 179128466.164]
Sec-HGS-Sched	**1065676446.564** [± **101692277.056**]	2143359732.256 [± 211454784.794]	4294563557.141 [± 373883906.206]	**8514397268.110** [± **602503134.100**]
Sec-IGA	1085575340.426 [± 110993632.105]	**2138208217.698** [± **221258711.190**]	**4236077792.436** [± **404456270.115**]	8593447951.179 [± 551754278.966]
Sec-(GA+TS)	1102225326.145 [± 197874153.696]	2199643747.642 [± 189693364.444]	4296055243.299 [± 386740590.285]	8608732539.636 [± 504003787.972]
		Dynamic Instances		
GA-SS-ANN	1163342728.245 [± 136548966.434]	2161846250.347 [± 272493690.708]	4322245472.632 [± 533180226.552]	8734534678.245 [± 635468708.749]
Sec-HGS-Sched	**1100334164.177** [± **181318391.192**]	**2113783653.774** [± **22486203.090**]	**4269654378.495** [± **547345211.754**]	8701108455.913 [± **779952031.937**]
Sec-IGA	1198943746.287 [± 139503645.521]	2198965387.563 [± 221434723.381]	4308567534.205 [± 50953994.605]	**8666386800.606** [± **884803516.367**]
Sec-(GA+TS)	1189239424.349 [± 132687197.083]	2197268532.324 [± 167974536.172]	4320061767.548 [± 468802204.277]	8800435684.376 [± 800470337.071]

Table 5.8 Average values of failure rate $Fail_r$ parameter for single-population, multi-level and hybrid genetic schedulers [$\pm s.d.$], ($s.d.$ = standard deviation)

Strategy	Small	Medium	Large	Very Large
		Static Instances		
GA-SS-ANN	3.993% [± 0.98]	4.089% [± 1.56]	8.436% [± 1.67]	8.736% [± 2.09]
Sec-HGS-Sched	**3.522%** [± **1.04**]	**4.011%** [± **1.33**]	5.342% [± 0.98]	**5.328%** [± **1.02**]
Sec-IGA	4.167% [± 0.98]	4.324% [± 1.26]	5.944% [± 1.89]	6.035% [± 2.23]
Sec-(GA+TS)	4.378% [± 0.92]	**5.016%** [± 1.05]	6.223% [± **1.35**]	6.927% [± 1.56]
		Dynamic Instances		
GA-SS-ANN	4.880% [± 0.98]	6.097% [± 1.62]	7.456% [± 1.32]	7.026% [± 2.11]
Sec-HGS-Sched	**3.116%** [± **0.80**]	**3.994%** [± **0.93**]	**4.250%** [± **0.99**]	**4.845** [± **1.14**]
Sec-IGA	4.030% [± 0.90]	4.951% [± 0.81]	5.016% [± 1.05]	5.136% [± 1.36]
Sec-(GA+TS)	4.93% [± 1.18]	6.93% [± 0.99]	6.327% [± 0.94]	6.001% [± 2.10]

Table 5.9 The number of genetic epochs necessary for the generation of the best solutions found by single-population, hybrid and multi-level schedulers

Strategy	Small	Medium	Large	Very Large
		Static Instances		
GA-SS-ANN	1722 (67.26%)	2733 (53.37%)	7992 (78.04%)	17739 (86.61%)
Sec-HGS-Sched	**1645 (64.25%)**	**2118 (41.24%)**	**6995 (63.60%)**	**15355 (74.97%)**
Sec-IGA	1788 (69.84%)	2475 (48.33%)	7044(68.78%)	16227 (79.22%)
Sec-(GA+TS)	1702 (66.48%)	2688 (52.50%)	7110 (69.42%)	15998 (78.11%)
		Dynamic Instances		
GA-SS-ANN	1522 (60.60%)	3021 (59.23%)	8010 (78.52%)	17830 (87.06%)
Sec-HGS-Sched	**1505 (59,93%)**	2935 (57.54%)	**7877 (77.22%)**	**16285 (79.51%)**
Sec-IGA	1612 (64.19%)	**2924 (57.33%)**	8054 (78.96%)	17922 (87.45%)
Sec-(GA+TS)	1578 (62.84%)	2989 (58.60%)	8035 (78.77%)	17911 (87.39%)

5.6 Conclusions

This chapter addressed the problem of the integration of the security mechanisms as additional criterion in the grid scheduling. Artificial Neural Network (ANN) was successfully implemented as the support mechanism for risk resilient genetic-based schedulers. The high effectiveness of this support was demonstrated by a comparison of the results of the performance of various GA-based schedulers in risky and secure scheduling scenarios. The proposed neural networks model seems to be a good solution for automatic monitoring of the grid system performance, but also a candidate technology for supporting the decision processes of the grid users and managers. In fact, all grid users working at different levels of the system, may specify their own strategies and preferences related to the security aspects in the scheduling process. In online scheduling, the users decisions are usually supported by the fuzzy-based online learning methodologies [110]. In batch scheduling the users' strategies and actions may be modelled by using the game-theory, as it is presented in the next chapter.

Chapter 6
Game-Theoretical Models of the Grid User Decisions in Security-Assured Scheduling: Basic Principles and Heuristic-Based Solutions

Abstract. This chapter presents two non-cooperative game approaches, namely the *symmetric non-zero sum* game and *asymmetric Stackelberg* game, for modelling the grid users' behavior. These models allow to illustrate new scenarios in scheduling and resource allocation problems, such as asymmetric users' relations, security and reliability restrictions in computational grids (CGs). Four GA-based hybrid schedulers are implemented for the approximation of the equilibrium states of both games. The proposed hybrid resolution methods are empirically evaluated through the grid simulator under the heterogeneity, security, large-scale and dynamics conditions.

6.1 Introduction

The security scheduling conditions defined in the previous chapter, may not be specified just by the analysis of the type of the applications submitted to grid, some local access policies to the grid resources, or the behavior and system security attributes defined in Sec. 5.3. Different types of the grid users may address their own individual requirements for the secure assignments of their tasks to the most trustful resources. In such a case the scheduling problem may be formulated as the decision problem of the grid users working at the different levels of the grid system.

In CGs the system management techniques must be able to group, predict, and classify different sets of rules, configuration directives, and environmental conditions. This management model must effectively deal with uncertainties in system information, that may be incomplete, imprecise, fragmentary, or overloading to control specific constituents and objects within intricate configurations. Decision scenarios ought to be outlined assuming partial visibility of environmental conditions, user heterogeneity, and resource dynamism in order to determine and select adequate evaluation criteria and assignment scores to render a final integrated decision result.

In many decision-making problems, most of the information is provided by humans, which is inherently non-numeric. Partial evaluations, preferences, weights are expressed linguistically. The evident role of fuzzy sets in decision-making and

J. Kołodziej: Evolutionary Hierarchical Multi-Criteria Metaheuristics, SCI 419, pp. 113–135.
springerlink.com

associated important processes such as consensus building, is well documented in the literature [118], [119], [69]. However, in large-scale grid systems all the users' information must be analyzed and interpreted in a short time, and special users' preferences must be taken into account. The fuzzy rules may be then wrongly interpreted by the system management components, or filtered in order to design the optimal strategies for minimizing the scheduling costs.

Game-theoretical models may be considered as alternative solutions for large-scale decision problems in highly parametrized heterogeneous environments. All scheduling criteria can be aggregated and defined as cumulative users' cost or pay-off functions, which makes the game models very useful in the analysis of the various users' strategies in the resource allocation process. Game-based models combined with the economic theory can capture many realistic scenarios in computational markets, computational auctions, grid and P2P systems as well as security and information markets. An important challenge in using game-theoretic models for grid scheduling and resource management is the large size scale of the grid system and the fact that resources cross different administrative domains. The grid game players should behave rationally, pursue well-defined objective functions (cost or pay-off functions), and react fast to the other players' actions and decisions.

This work is based on the results presented in [85], where the preliminary versions of the symmetric and asymmetric non-cooperative grid users games were defined for the purpose of illustrating the users strategies in the security-aware scheduling. This chapter summarizes those results and presents the formal models of general symmetric and asymmetric Stackelbelg games. These models are based on the premise about the users' behavior in a realistic large-scale grid, where users, usually independently of each to another, submit their tasks/applications to the grid system. Additionally, in the Stackelberg game, one player (user) is acting as a Leader with a privileged access to resources. This Leader assigns his tasks first, and the rest of the users (Followers) react rationally to the Leader's actions. The Followers do not cooperate with each other, but their decisions depend on the Leader's action. This model illustrates very well the real-life situation, where the roles of the users are in fact asymmetric with regard to their access rights and usage of resources. It must also be noted that in many economical models the sellers and buyers stand in asymmetric positions as well. Having a control over large resource pools and maintaining the large fraction of the task batch for scheduling can be the reasons of having some privileges in the resource access, or in the setting of the reasonable resource utilization pricing policies.

The users cost functions in the game are interpreted as the cummulative cost of the secure execution of their tasks and the costs of the utilization of resources. The cumulative cost function specified for the whole game is optimized at global and local (users') levels, through four genetic-based hybrid meta-heuristics, which combine Genetic Algorithm (GAs) and modified Minimum Completion Time (MCT) method.

Two scheduling scenarios are considered in in this work, namely *risky* and *secure* mode. In the former, security conditions are ignored by the users by allocating their tasks to all available machines, independently of the trusted levels. In the later, users allocate tasks to available machines assuring task security demands. It should be noted that in task scheduling the definition of the security demand can be two-fold: (a) tasks can have security demands on resources to be allocated at and (b) resources can have security demands on tasks to be assigned to them. This work is focused on the condition (a) of security requirements.

The proposed models were evaluated under the heterogeneity, the large scale and dynamics conditions using the *Sim-G-Batch simulator*. The relative performance of four hybrid schedulers are measured by the makespan and flowtime metrics. However, the main aim of the empirical analysis is to compare the effectiveness of the game models in the reduction of the scheduling costs in the secure scenario, and the results achieved by the best single-population grid scheduler generated in Chapter 5 for the secure scheduling with the ANN support (see Sec. 5.5.3.1).

6.2 Users' Behavior Models in Grid Scheduling

The classifications of the grid scheduling problems presented in Sec. 1.3.2 in Chapter 1 do not span over the analysis of the relations and behavior of the grid users at different systems levels (see Fig. 1.3). Three basic models of grid users' relation in grid scheduling processes can be defined as follows:

- **Cooperativeness:** In this case the users can form a coalition to plan in advance their actions;
- **Non-cooperativeness:** In this scenario the users act independently of one another;
- **Semi-cooperativeness:** In this model each user can select a partner for the cooperation.

The analysis of the above mentioned relations of the users is used for the specification of the generic models of the following three types of grid user games, namely *non-cooperative*, *cooperative* and *semi-cooperative* games:

- In **non-cooperative game** the players act independently of each other. This model is based on the premise about the users' behavior in realistic grids, where cooperation is difficult in large-scale system, and grid users submit their tasks independently. Also the resource owners act selfishly in order to maximize the resource utilization and to execute the tasks from the local users.
- In **cooperative game** the players can form a coalition to plan their future actions. This model is useful for the intra-site grid negotiations, where the local job dispatchers can define the joint "execution capabilities" parameters for the clusters of the grid sites and declare them to the global scheduler.
- In **semi-cooperative game** each player can choose (randomly) another player for cooperation. This game is usually proposed as a multi-round auction to incorporate the task rescheduling.

The solution of each of those games is an equilibrium state, in which each player holds correct expectations concerning the other players' behavior (see [75] for the detailed analysis).

The users can have different privileges to the resources, resulting in the examination of the following two scenarios:

- **Symmetric scenario.** In this case there are no special privileges in the resource usage for the grid users.
- **Asymmetric scenario.** In this case there is a privileged user (Leader), who can have full access to resources as opposed to the rest of users who can be granted only limited access to resources. The Leader could also be the owner of a large portion of the task pool, as it is reasonable to allocate first his tasks at best resources in the system.

The game models presented in this chapter are based on the non-cooperative scenario in symmetric and asymmetric modes.

6.3 Symmetric and Asymmetric Games of Independent Grid Users

One of the main benefits of the game-based scheduling and resource management in CGs is that it enables a scalability and personalization of the decision-making processes of grid users and resource owners. Due to the sheer scale of grid systems, the non-cooperative game is a potential model for integrating security and resource reliability requirements in grid scheduling. This section presents two different general scenarios of the non-cooperative grid users behaviors, namely *symmetric* and *asymmetric* strategic game models.

6.3.1 Non-cooperative Symmetric Game

Let us denote by $Play$ the number of grid users (players). The total number of tasks $n \in N$ in a given batch can be expressed as the sum of numbers of tasks submitted by all users, i.e.

$$n = \sum_{a=1}^{Play} k_a, \tag{6.1}$$

where k_a is the number of tasks of the user $a = 1, \ldots, Play$.

Each player a controls his strategic variables defined as the following *user's strategy vector* :

$$Pl_a = \left[j_{(\widehat{k_{(a-1)}}+1)}, \ldots, j_{(\widehat{k_{(a-1)}}+k_a)} \right] \tag{6.2}$$

where $\widehat{k_{(a-1)}} = k_1 + \ldots + k_{(a-1)}$

The schedules can be then expressed by the following vectors of the users' parameters:

$$S = \left[i_1^1, \ldots, i_{k_1}^1, \ldots, i^a_{(\widehat{k_{(a-1)}}+1)}, \ldots, i^a_{(\widehat{k_{(a-1)}}+k_a)}, \ldots \ldots i^{Play}_{k_{Play}} \right], \tag{6.3}$$

in the direct representation, and

$$Sch = [Pl_1, \ldots, Pl_{Play}], \tag{6.4}$$

in permutation-based representation (see Chapter 2, Sec. 2.2.1 for details on the schedules representations).

In *symmetric* non-cooperative users' game the privileges to the resources are the same for all users. Each user tries to choose an optimal strategy for the assignment of his tasks to machines in order to minimize his cost of tasks scheduling and, as the results, also the overall scheduling costs. An illustrative example of the symmetric game can be a scheduling scenario in which each player submits an equal amount of tasks, i.e. $k = k_1 = k_2 =, \ldots, = k_{Play}$. It means that in such a case the total number of tasks in the batch can be calculated in the following way: $n = Play \cdot k$.

Definition 6.1. The symmetric grid users' non-cooperative game can be defined as a tuple
$G_{Play} = (Play; \{J_a\}_{a=1,\ldots Play}; \{Q_a\}_{a=1,\ldots,Play})$, where:

- $Play$ is the number of grid users;
- $\{J_1, \ldots, J_{Play}\}$; are the sets of users' strategies;
- $\{Q_1, \ldots, Q_{Play}\}; Q_a : J_1 \times \ldots \times J_{Play} \to \mathbb{R}; \forall_{a=1,\ldots,Play}$ is the set of users' cost functions.

The users' strategy vectors Pl_a are the elements of the strategy spaces $J_l = J_{((a-1)\cdot k+1)} \times \ldots \times J_{(a\cdot k)}$ specified for each user $a, (a = 1, \ldots, Play)$. The cost of playing the game calculated for a particular user a is defined as the cost of scheduling of his tasks (or the user's cost function) and is denoted by Q_a. The players try to minimize simultaneously their cost functions Q_a in the game.

Definition 6.2. A multi-vector $(\widehat{Pl_1}, \ldots, \widehat{Pl_{Play}})$ of strategies is called **an equilibrium state (point)** of the game if :

$$\begin{aligned} & Q_a\left(\widehat{PL_1}, \ldots, \widehat{PL_{Play}}\right) = \\ & = \min_{Pl_1 \in J_1} Q_a\left(\widehat{Pl_1}, \ldots, \widehat{PL_{(a-1)}}, PL_a, \widehat{Pl_{(a+1)}}, \ldots, \widehat{Pl_{Play}}\right) \end{aligned} \tag{6.5}$$

for all $a = 1, \ldots, Play$.

The formulas of Q_a functions are specified for the permutation-based representation of the schedules because of the simpler notation used in the Eq. (6.4). Those procedures can be easily transformed into the direct representation by substituting the $\left(\widehat{PL_1}, \ldots, \widehat{PL_{Play}}\right)$ vector by the vector defined in Eq. (6.3). The equilibrium point can be interpreted as a steady state of the a strategic game, in which each player holds correct expectations concerning the other players behavior[1]. If the strategies

[1] In the case of continuous players' cost functions the equilibrium state of the game is called the *Nash equilibrium* [39].

chosen by all players are equilibrium points, no player is interested in changing his strategy.

To be a solution of the grid users', the game the equilibrium point should be additionally Pareto-optimal [139, 117]. In this chapter we consider the non-zero sum games[2], for which the equilibrium points are the results of minimization of a *multi-cost game function* Q defined as follows.

Let us denote by $minQ_a, (a = 1, \ldots Play)$, the minimal value of the function Q_a calculated for each user a, that is to say:

$$minQ_a = \min_{PL_a \in J_a} \{Q_a(PL_1, ..., PL_{Play})\}. \tag{6.6}$$

The results of the global minimization of the following *game multi-cost* function $Q : J_1 \times \cdots \times J_{Play} \to \mathbb{R}$:

$$Q(PL_1, ..., PL_{Play}) = \sum_{a=1}^{Play} \frac{1}{Play} \left(Q_a(PL_1, ..., PL_{Play}) - minQ_a\right), \tag{6.7}$$

is an equilibrium state of non-cooperative non-zero sum symmetric game of the grid users, which satisfies the condition of the Pareto-optimality [117][3].

6.3.2 *Asymmetric Scenario – Stackelberg Game*

The symmetric games are quite simple for the implementation and well studied for many high-performance computing approaches. However, the symmetric scenario may not be a good model of the realistic users' relations. Due to the cross-domain access, authorization and resource management features of the grid system, the grid users have different access policies to the resources and they stand in an asymmetric position with regard to resource usage privileges. The asymmetric behavior of grid users directly impacts the results of the scheduling process.

A Stackelberg game is the simplest model for illustrating the asymmetric scenario of the behavior of non-cooperative grid users. In this game one privileged user acts as a *Leader*, and the rest of players (users) are his *Followers*.

The Stackelberg games have been well-studied in the game theory literature (see, e.g. [10]). Roughgarden [126] defined a Stackelberg game model for scheduling tasks on a set of machines with load-dependent latencies in order to minimize the total latency of the system.

The following examples illustrate some real-life grid scenarios, to which the Stackelberg game model can be applied:

---

[2] In this scenario the strategies of the players are not opposite, i.e. the sum or the values of all players cost functions Q_a is not 0.

[3] In fact, the function Q is a special case of the weighted or weighted distance L^p metric function with $p = 1$. The values of all Q_a functions are non-negative, and the weight coordinates are strictly positive, which means that the global solutions of the problem defined in Eq. (6.7)are Pareto-optimal [142].

- There is a privileged grid user (Leader), who can have the full access to resources as opposed to the other users with limited access to resources.
- Some tasks can have critical deadlines (especially in online scheduling) and they can be sent by the Leader to the meta-broker with a request to allocate them first.
- Considering a batch of tasks, the Leader can be the owner of a large portion of the tasks in the batch; therefore it might be reasonable to allocate his tasks to the best resources in the system.
- Some tasks could have security requirements. Therefore the Leader can send such an information and all security parameters to the scheduler or directly to the meta-broker requesting to allocate them in the most trustful resources (secure machines).
- Tasks submitting to a grid system could be varied in their needs for computational resources. Some of them could be atomic tasks generated by compound tasks while the others could be just monolithic applications. The high degree of heterogeneity of tasks usually has a great impact on the grid system's performance. In such a scenario the Leader could create a small batch of the most time consuming tasks as the backlog set of grid applications, in order to "balance" the computational loads of machines during the scheduling. These tasks would be sent to the meta-broker requesting to allocate them first.

Formally the two-level Stackelberg game of the grid users can be defined in the following way:

- **Leader's Level: Leader's action I -** Leader chooses his initial strategy $\widetilde{Pl_1} = [\widetilde{j_1}, \ldots, \widetilde{j_{k_1}}]$, where k_1 denotes the number of tasks submitted by the Leader.
- **Followers' Level: Followers' action -** Followers minimize simultaneously their cost functions relative to the Leader's choice:

$$\begin{cases} Pl_2^{Fol} = arg\min_{(Pl_2 \in J_2)} \{Q_2(\widetilde{Pl_1}, Pl_2, \ldots, Pl_{Play})\} \\ \vdots \\ Pl_{Play}^{Fol} = arg\min_{(Pl_{Play}) \in J_{Play}} \{Q_{Play}(\widetilde{Pl_1}, \ldots, Pl_{Play})\} \end{cases} \quad (6.8)$$

  where J_1 is the set of the Leader's strategies and Q_a is the cost function of the user a defined as in the symmetric case in Eq. (6.12). Let us denote by $Pl^{Fol} = \left[\widetilde{Pl_1}, Pl_2^{Fol}, \ldots, Pl_{Play}^{Fol}\right]$ a *Followers' Vector*, which is interpreted as the result of the Followers' action.
- **Leader's Level: Leader's action II -** Leader updates his strategy by minimizing his cost function Q_1 (see also Eq. (6.12)) taking into account the result of Followers' actions. The following vector $Pl^G = \left[Pl^{Lead}, Pl_2^{Fol}, \ldots, Pl_{Play}^{Fol}\right]$, where:

$$Pl^{Lead} = arg \min_{(Pl_1 \in J_1)} Q_1 \left(Pl_1, Pl_2^{Fol}, \ldots, Pl_{Play}^{Fol}\right) \quad (6.9)$$

  is a solution of the whole game.

It has to be noted that the Followers can play an "ordinary" non-cooperative symmetric game, but they must know the Leader's action first. The game multi-cost function Q in this case can be defined in the following way:

$$Q_{Stac} = \frac{1}{Play} Q_1 + Q_{Fol}; \tag{6.10}$$

where Q_1 is the Leader's cost function and

$$Q_{Fol} := \frac{Play - 1}{Play} \sum_{a=2}^{Play} Q_a; \tag{6.11}$$

is a *Followers' cost function*. An optimal solution of the whole game is called *Stackelberg Equilibrium*.

6.3.2.1 Users' Cost Functions in Security-Aware Scheduling

In conventional grid scheduling with the typical scheduling objective functions such as makespan and flowtime, the users' costs of scheduling their tasks are limited to the costs of tasks execution. In utility grids, there are the resource utility functions that must be specified for the calculation of the resource utilization cost [50]. In security-assured scheduling some additional costs must be considered. The users have to "pay" an additional "fee" for the secure allocation of their tasks in the machines. In this work all of those costs are integrated into cumulative-cost functions $Q_a, a \in \{1, \ldots, Play\}$, defined separately for each grid user as the weighed sum of the following three components:

$$Q_a(S) = Q_a^{(ex)}(S) + Q_a^{(util)}(S) + Q_a^{(sec)}(S), \tag{6.12}$$

where:

- $Q_a^{(ex)}$ indicates the user's task execution cost ,
- $Q_a^{(util)}$ denotes the resource utilization cost , and
- $Q_a^{(sec)}$ is the cost of security-assured allocation of the user tasks .

In this work the ETC Matrix model (see Chapter 2) is used for the specification of all cost functions for the users.

6.3.3 *Task Execution Cost*

The total cost of execution of the user's tasks can be calculated as an average completion time of his tasks on machines, to which they are allocated[4]. In terms of the

[4] The values of all components of the users game cost functions, i.e. $Q_a^{(ex)}$, $Q_a^{(util)}$ and $Q_a^{(sec)}$ functions, are scaled to get the all values in the same range.

completion times of machines (see Chapter 2, Eq. (2.13)) the function $Q_a^{(ex)}$ can be defined using the following formula:

$$Q_a^{(ex)} = \frac{\sum_{j=k_{a-1}+1}^{k_a} completion[j][i]}{completion_m \cdot k_a}, \tag{6.13}$$

where $completions[j][i]$ denotes the completion time of a task j on a given machine machine i and and it is calculated in the following way:

$$completions[j][i] = ETC[j][i] + ready[i]. \tag{6.14}$$

In Eq. (6.13), the $completion_m$ indicates the maximal completion time of all tasks submitted by the user a, that is to say:

$$completion_m = \max_{j=k_{a-1}+1,\ldots,k_a} completion[j][i]. \tag{6.15}$$

6.3.4 Resource Utilization Cost

The grid user's utility function is usually defined as a cost of buying free CPU cycles [50]. In this work the utilization cost paid by the user a is calculated as a "portion" of the average idle time of machines on which his tasks are executed. This cost depends on the completion times of the user's tasks. The utility function $Q_a^{(util)}$ is defined as follows:

$$Q_a^{(util)} = \sum_{i \in machines(a)} \left(1 - \frac{Completion_{(a)}[i]}{C_{max}}\right) \cdot Idle\_Factor[i] \tag{6.16}$$

where $machines(a)$ denotes the set of machines, to which all tasks of the user a are assigned and C_{max} refers to the makespan. The completion time of a given machine $i \in machines(a)$, denoted by $Completion_{(a)}[i]$, is calculated in the following way:

$$Completion_{(a)}[i] = ready[i] + \sum_{\substack{j \in N: \\ S[j]=i}} ETC[j][i] \tag{6.17}$$

where $S[j]$ is the value of j-th coordinate in a given schedule vector S (or Sch – both implementations of the schedules may be used in Eq. (6.17). The following expression:

$$\left(1 - \frac{Completion_{(a)}[i]}{C_{max}}\right) \cdot Idle\_Factor[i], \tag{6.18}$$

in Eq. (6.16) is interpreted as an idle time of machine i calculated for a given user a. This is just a "portion" of the total idle time of machine i, and it is proportional to the time of execution of all tasks of the user a assigned to this machine. This proportion is specified by the coefficient $Idle\_Factor[i]$ in the following way:

$$Idle\_Factor[i] = \frac{\sum_{j \in Tasks_{(a)}[i]} ETC[j][i]}{Completion_{(a)}[i]} \tag{6.19}$$

where $Tasks_{(a)}[i]$ is the set of the tasks of the user a assigned to the machine i.

It follows from Eq. (6.16) that the utilization cost is minimal in the case of allocation of the user tasks to machines with the maximal completion times.

6.3.5 Security-Assurance Cost

The security-assurance cost of scheduling the tasks of the user a, denoted by $Q_a^{(sec)}$ in Eq. (6.12, depends on the scheduling strategy and the result of the verification of security condition by the trust manager. The manager must analyze the security demand SD and trust level TL vectors for tasks and machines and the *Machine Failure Probability* matrix $Pr_f = [Pr_f[j][i]]_{n \times m}$ must be specified in the similar way as in Eq. (5.2), that is to say:

$$Pr_f[j][i] = \begin{cases} 0 & , sd_j \leq tl_i \\ 1 - e^{-\alpha(sd_j - tl_i)} & , sd_j > tl_i \end{cases} \tag{6.20}$$

where α is the failure coefficient and sd_j and tl_i are the security demand and trust level parameters for task j and machine i.

Similarly to Sec. 5.3.1 in Chapter 5, two different scheduling strategies can be considered, namely *Risky* and *Secure* modes, in order to illustrate the various users' and grid managers' decisions. The formulas for calculating the security cost for the users are based on the formulas for the completion times of machines and flowtime in Eq. (5.3)– (5.6).

In the **Risky mode** all risky and failing conditions are ignored by the users. In this case the "security" components of the functions Q_a are in fact not calculated, i.e. $Q_a^{(sec)} = 0, \forall a = 1, \ldots, Play$. However, some machines may fail during the tasks execution because of too restrictive security requirements and rescheduling procedure of those tasks must be activated. Therefore the security cost in this case is calculated as follows:

$$Q_l^{(sec)}[ris] = \sum_{j \in Res(a)} \frac{P_f[j][i] \cdot ETC[j][i]}{(ETC)_{m(a)} \cdot \sharp(Res(a))}, \tag{6.21}$$

where $Res(a)$ is the set of the tasks of the user a which must be rescheduled, and $(ETC)_{m(a)}$ is the (expected) maximal computation time of the tasks of the user a in a considered schedule, i.e.:

$$(ETC)_{m(a)} = \max_{\substack{j \in Task(a) \\ i \in machines(a)}} ETC[j][i]. \tag{6.22}$$

In Eq. (6.22), $Task(a)$ denotes the set of the tasks of the user a and $machines(a)$ is the set of the machines to which the user tasks are mapped in a considered schedule.

In **Secure mode** the users must pay the cost of the verification of the security condition for his tasks (see Sec. 5.3). The cost of possible failures of machines during the tasks executions are calculated as the products of the failure probabilities and the expected times of computation of the tasks on the inaccessible machines. The secure cost function $Q_a^{(sec)}$ in this case is defined as follows:

$$Q_a^{(sec)}[secure] = \sum_{j=k_{a-1}+1}^{k_a} \frac{P_f[j][i] \cdot ETC[j][i]}{(ETC)_{m(a)} \cdot k_a}, \tag{6.23}$$

The security-assurance cost expressed as $Q_a^{(sec)}[secure]$ for each grid user is minimized. It means that each user tries to allocate his tasks in the most trustful resources and the values of task failure probabilities $P_f[j][i]$ should be minimal.

6.4 Solving the Grid Users Games

The problem of solving the finite strategic game remains challenging especially in real-life approaches. In order to compute the values of the game cost functions Q defined in Eqs. (6.7) and (6.10), the cost functions of all players must be first minimized. Therefore the problem of the minimization of Q function can be defined as a hierarchical procedure presented in Fig. 6.1. This procedure is composed of two cooperating modules: **Global Module**, in which the values of the function Q are calculated and optimized, and the **Players' Module** - which solves the local level problems of the minimization of the users' cost functions Q_a.

The communication procedure between *Global* and *Players' Modules* can be defined as follows: Let us denote by $S^{(0)}$ an initial schedule generated in the Global Module, i.e. $S^{(0)} = [Pl_1^{(0)}, \ldots, Pl_{Play}^{(0)}]$, where $Pl_a^{(0)}$ is the initial strategy vector of the

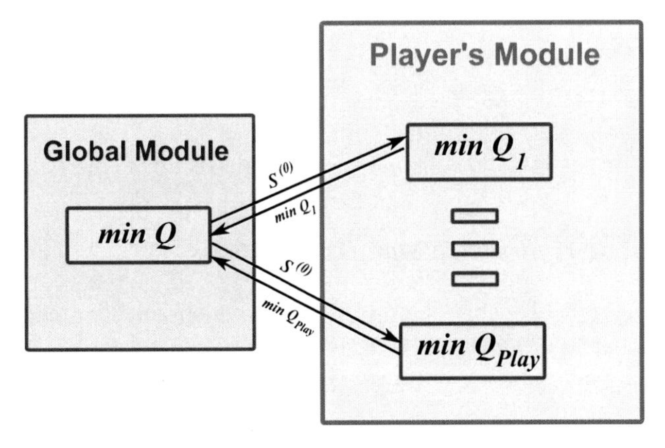

Fig. 6.1 Hierarchical procedure of solving non-cooperative symmetric game of grid users

user a (see Eq. (6.2)). Vector $S^{(0)}$ is replicated and its copies are sent to the *Players' Module* - one copy per user. Then, each user independently optimizes his game cost function[5] by changing the allocations of just his own tasks. As the result of this minimization, the optimal values of the Q_a cost functions are calculated:

$$\begin{cases} minQ_a{}^{(0)} = \min_{(Pl_1 \in J_1)} Q_1 \left(Pl_1, Pl_2^{(0)}, \ldots, Pl_{Play}^{(0)} \right) \\ \vdots \\ minQ_{Play}{}^{(0)} = \min_{(Pl_{Play} \in J_{Play})} Q_{Play} \left(Pl_1^{(0)}, \ldots, Pl_{Play-1}^{(0)}, Pl_{Play} \right) \end{cases} \tag{6.24}$$

These values are sent back to the **Global Module**, where the objective function for the whole game Q is calculated for the schedule $S^{(0)}$.

In the case of Stackelberg game the *Global Module* plays the role of the *Leader's* component and the *Player's Module* – the *Follower's* procedure, as it is presented in Fig. 6.2.

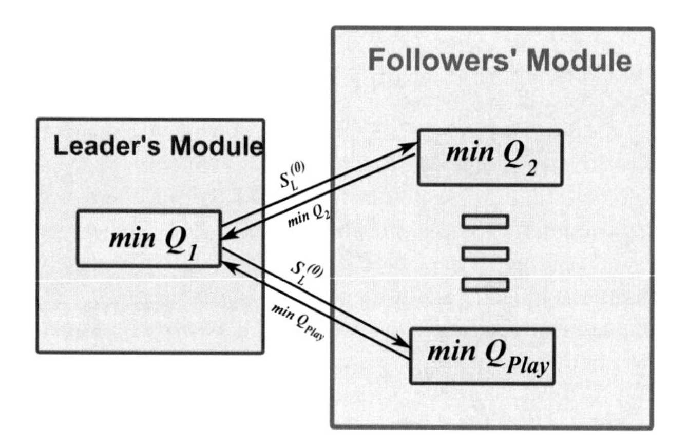

Fig. 6.2 Hierarchical procedure of solving Stackelberg game

However, in this case the schedule vector $S_L^{(0)}$ before its replication and sending to the Followers, is partially 'fulfilled' by the Leader. The Leader makes his preliminary assignments and send the incomplete schedule vectors to the Followers.

6.4.1 Genetic Hybrid Metaheuristic Solvers

Similar to the empirical analysis presented in the previous chapters, and due to multiple constraints and different preferences of the grid users, genetic-based heuristic approaches seem to be the best candidate methodologies for solving the users'

[5] Note that the users costs optimization in the **Players' Module** can be implemented as a parallel multi-threaded procedure, which can speed-up the whole process.

games. However, in this case the main framework of the scheduler must be extended by the hybridization of genetic algorithm (GA) working in the **Global/Leader's Module** with some other heuristic method implemented in the **Players'/Followers' Module**.

Four hybrid GA-based schedulers have been defined for solving the symmetric and asymmetric grid users' games. The combinations of the heuristic components of these hybrids are presented in Table 6.1.

Table 6.1 Hybrid meta-heuristics for risky and secure-assured scheduling

Meta-heuristic	**Global/Leader's Module**	**Players'/Followers' Module**
RGA-GA	RGA	PGA
RGA-PMCT	RGA	PMCT
SGA-GA	SGA	PGA
SGA-PMCT	SGA	PMCT

The GA-based meta-heuristics may work as global and local optimizers in the **Global/Leader** and **Players'/Followers'** Modules. Each hybrid algorithm is defined as a combination of two methods, namely *Risky Genetic Algorithm (RGA)* and *Secure Genetic Algorithm (SGA)*– in the **Global Module**; and two local level optimizers, namely *Player's Genetic Algorithm (PGA)* and *Player's Minimum Completion Time (PMCT)*– in the **Players' Module**. It can be observed that, in fact, it is not necessary to replicate the whole population form the Global or Leader's Module to the Players'/Followers' Module. For each player independently, just the changes in machine completion times must be updated. Therefore the general procedure of Players'/Followers algorithm may be defined as follows:

Algorithm 3. The optimization procedure in **Player's/Followers' Module**

1: Send the *ready_times* vectors to the individual players;
2: Individual players compute the $MinQ_a$ values;
3: Receive the $MinQ_a$ values from the individual players;
4: Send the $MinQ_a$ values to **Global Module**;

Schedulers Implemented in Global and Leader's Module

The generic template of the main GA-based engine of the hybrid scheduler designed for solving the symmetric games is presented in Alg. 4. This template is similar to the Alg. 1 defined in Chapter 3 for HGS-Sched (see. Sec. 3.3).

Algorithm 4. Genetic Algorithm template

1: Generate the initial population P^0 of size μ;
2: Send the *ready_times* vectors of the machines corresponding to the individuals of the population P^0 to the **Player's Module**;
3: Receive the $minQ_a$ values from the subordinate unit
4: Evaluate P^0;
5: **while** not termination-condition **do**
6: Select the parental pool T^t of size λ; $T^t := Select(P^t)$;
7: Perform crossover procedure on pars of individuals in T^t with probability p_c; $P_c^t := Cross(T^t)$;
8: Perform mutation procedure on individuals in P_c^t with probability p_m; $P_m^t := Mutate(P_c^t)$;
9: Send the *ready_times* vectors of the machines corresponding to the individuals of the population P_m^t to the **Player's Module**;
10: Receive the $minQ_a$ values from the Players' Module
11: Evaluate P_m^t ;
12: Create a new population P^{t+1} of size μ from individuals in P^t and/or P_m^t ;
13: $t := t + 1$;
14: **end while**
15: **return** Best found individual as solution;

The main difference between *RGA* and *SGA* algorithms is the method of the evaluation of the population by using the users' cost functions Q_a, which is different in the Risky (RGA) and Secure (SGA) modes. The formulas of calculating the 'security' costs in both scenarios are defined in Sec. 6.3.5.

In the case of Stackelberg game the initialization procedure in the main GA algorithm is a bit different than in the symmetric scenario. The general template of the main genetic engine at the **Leader's (Global) Module** in this game is defined in Alg. 5.

Algorithm 5. A GA-based scheduler at the **Leader's** level

1: Generate P^0 containing μ "incomplete" schedules; $t = 0$;
2: Send P^0 to the **Followers** to complete the respective parts of all schedules in P^0; $P^0(F)$ is created;
3: Update the population P^0 according to the Followers' solutions;$P^0 := P^0(F)$;
4: Evaluate P^0;
5: **while** not termination-condition **do**
6: Select the parental pool T^t of size λ; $T^t := Select(P^t)$;
7: Perform crossover procedures separately on Leader's and Followers'variables on pairs of individuals in $T^t(F)$ with probability p_c; $P_c^t := Cross(T^t)$;
8: Perform mutation procedures separately to Leader's and Followers' variables on individuals in P_c^t with probability p_m; $P_m^t := Mutate(P_c^t)$;
9: Evaluate P_m^t ;
10: Create a new population P^{t+1} of size μ from individuals in P^t and P_m^t ; $P^{t+1} := Replace(P^t; P_m^t)$
11: $t := t + 1$;
12: **end while**
13: **return** Best found individual as solution;

The process of initialization of the population in the main GA algorithm is defined as a two-step procedure. In the first step, the P^0 set is generated as a candidate initial population. It consists of the incomplete schedules generated by the Leader by using one of the initialization methods for GA-based schedulers (see Chapter 3, Sec. 3.3). Each schedule from this set contains just the values of the Leader's decision variables. All those "incomplete" chromosomes are sent to the Followers' Module. The Followers complete each schedule by using one of the ad-hoc heuristics. The updated population P^0 is then evaluated under the game cost function Q_{Stac} defined in Eq. (6.10. The crossover and mutation operations are performed separately on Leader's and Followers' decision variables. Therefore in each generation the Followers can update their own decisions (including the initial choices) according to all changes in availability of resources introduced by the Leader.

Local Schedulers in Players' and Followers' Modules

Two modifications of well-known grid schedulers are implemented in the Players' and Followers' Modules.

The first scheduler, called *Player's Genetic Algorithm*, is a simple extension of the classical GA-based scheduler defined in Alg. 1 applied independently for each user with the cost function Q_a as the fitness measure. The genetic operations are executed on sub-schedules of the length k_a labeled just by the tasks submitted by user a. In the implementation presented in this work the GA procedures in the Players' or Followers Modules are executed sequentially for the "queue" of users, however each algorithm may be implemented as a separate process on parallel multiprocessor machine the number of processors must be in this case the same as the number of players or followers)

The second method, called *Player's Minimum Completion Time - (PMCT)*, is the modification of *Minimum Completion Time - MCT*. In this method, a task is assigned to the machine yielding the earliest completion time (defined as the sum of *ready_time* for the machine and time of computing all tasks assigned there). The process is repeated until there remain tasks to be assigned. The template of the main mechanism of *PMCT* procedure is defined in Alg. 6.

Algorithm 6. *PMCT* algorithm template

1: Receive the population of schedules and *ready_times* of the machines from the **Global Module**;
2: **for all** Schedule in the population **do**
3:    Calculate the completion times of the machines in a given schedule;
4:    **for all** Individual user/Follower **do**
5:       **for all** User's Task/Follower's Task **do**
6:          Find the machine that gives minimum completion time;
7:          Assign task to its best machine;
8:          Update the machine completion time;
9:       **end for**
10:       Calculate the $minQ_a$ value for a given schedule;
11:    **end for**
12:    Send the $minQ_a$ values to the **Global Module**;
13: **end for**

6.5 Empirical Analysis

The main aim of the empirical evaluation of the genetic hybrid schedulers defined in the previous section is to compare the effectiveness of the game-based models in the optimization of the main scheduling objective functions, namely *Makespan* and *Mean_Flowtime* defined in Sec. 5.5.2, with the results achieved by the best single population $GA-CX-R-ANN$ scheduler supported by the neural network mechanism in the similar analysis in Chapter 5.

Four hybrid meta-heuristics defined in Table 6.1 have been used for solving the symmetric and asymmetric games. These methods were integrated with the *Sim-G-Batch* simulator. The experiments have been conducted on two benchmarks composed by a set of static and dynamic instances. Similarly to the empirical analysis provided in Chapters 4 and 5, four grid size scenarios are considered, namely Small, Medium, Large and Very Large grids.

The key parameters of the simulator in all experiments are the same as in Table 5.1 The parameters of *HGS-Sched* for generating the GA algorithms in **Global**, **Leader's**, **Players'** and **Followers' Modules** are defined in Table 6.2.

Table 6.2 GA settings in the **Global/Leader's** and **Players'/Followers' Modules** for large static and dynamic benchmarks

Parameter	**Global/Leader's Module**	**Players'/Followers' Module**
period_of_metaepoch	$(5*(n))/10$	$(\lceil 0.5*(n)\rceil)/10$
nb_of_metaepochs	10	
population size (*pop_size*)	60	20
intermediate pop.	48	14
selection method	LinearRanking	
crossover method	CX	
cross probab.	0.8	0.8
mutation method	Rebalancing	
mutation probab.	0.2	
initialization	LJFR-SJFR + Random	
max_time_to_spend	500 secs (*static*) / 800 secs (*dynamic*)	

There are 16 players in symmetric game and 15 Followers in the Stackelberg game, and the number of the Leader's tasks is a half of the whole task batch. The coefficients of SD and TL vectors, and the machines reliability probabilities P_i are defined as the uniformly generated fractions in the ranges [0.6;0.9], [0.3;1] and [0.85;1] respectively. The value of the failure coefficient λ is 3.

Each experiment was repeated 30 times under the same configuration of parameters and operators.

The histograms of the average values of *Makespan* and *Mean_Flowtime* achieved by four hybrid meta-heuristics designed for solving the users games are presented in Fig. 6.3 and 6.3.

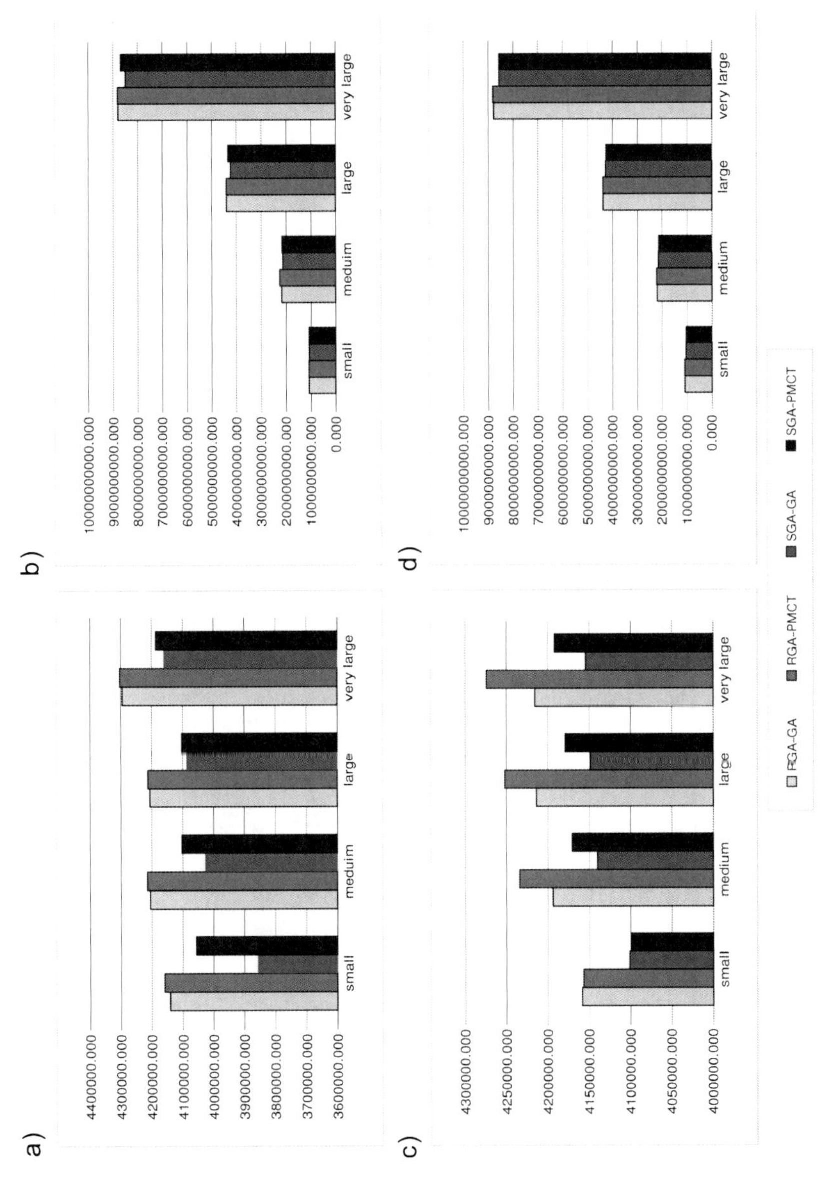

Fig. 6.3 Experimental results for non-cooperative symmetric game: in static case - (a) average *Makespan*, (b) average *Mean_Flowtime* ; in dynamic case - (c) average *Makespan*, (d) average *Mean_Flowtime*.

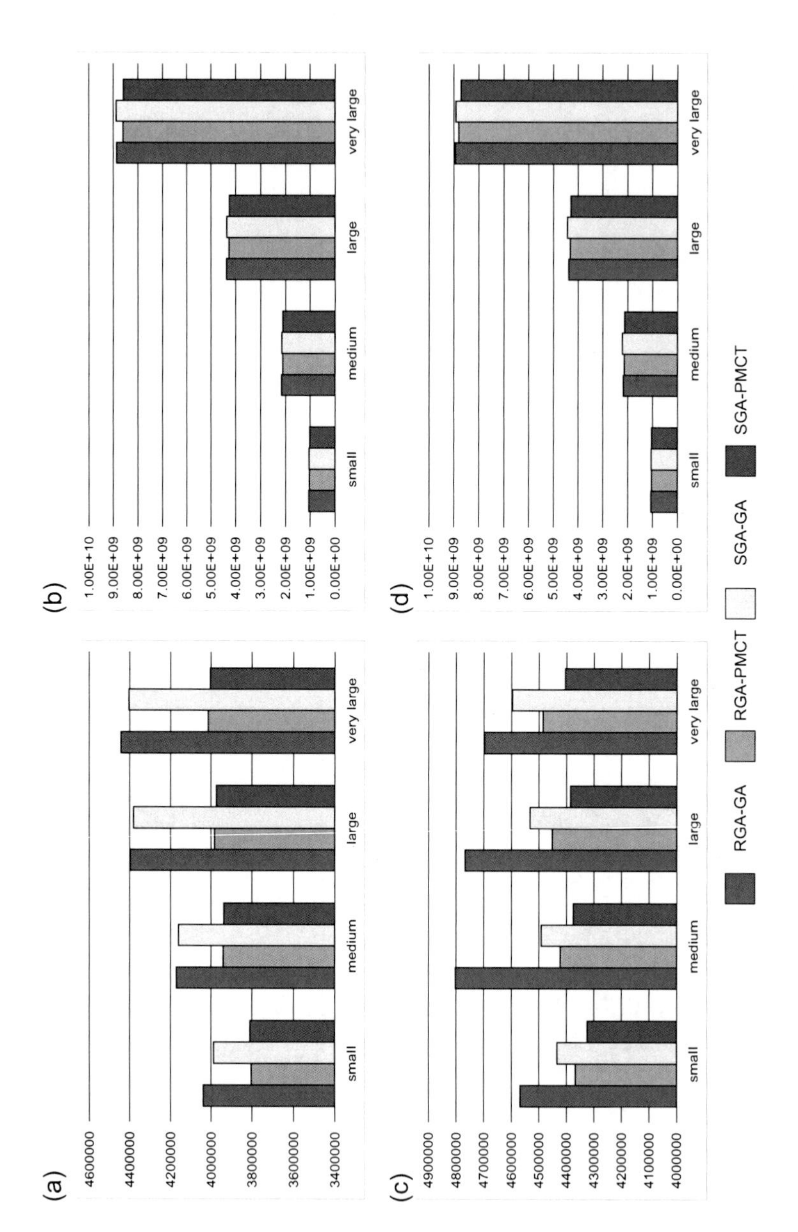

Fig. 6.4 Experimental results for Stackelberg game: in static case - (a) average *Makespan*, (b) average *Mean_Flowtime* ; in dynamic case - (c) average *Makespan*, (d) average *Mean_Flowtime*.

In the symmetric game the best results for *Makespan* and *Mean_Flowtime* in all considered grid scenarios were achieved by *SGA-GA* scheduler. Especially in static 'Small' grid this method is very effective in *Makespan* reduction. The differences in the *Mean_Flowtime* results achieved by all hybrid meta-heuristics are not so significant, while in the case of *Makespan* both *SGA* hybrids significantly outperform **risky** hybrids in all grid scenarios.

In the case of Stackelbeg game two *PMCT* hybrids outperform the *RGA-GA* and *SGA-GA* algorithms. For *Makespan* values the differences in the results achieved by *PMCT* and *GA* hybrids are significant, while in the case of *Mean_Flowtime* all values are at the same level, except those obtained for 'Very Large' grid size. The best results in all instances are achieved by *SGA-PMCT* algorithm. However, in the case of static scheduling the efficiencies of *RGA-PMCT* and *SGA-PMCT* are very similar, while in the dynamic case, especially for *Makespan* values, the differences in both schedulers performances are significant.

The results achieved by two most efficient meta-heuristics in optimizing the users' game costs, namely $SGA - GA$ in the symmetric game and $SGA - PMCT$ in Stackelberg game, have been compared with the results generated by the best single-population security-aware scheduler from the previous chapter, namely $GA - SS - ANN$ algorithm. Tables 6.3, 6.4 and 6.5 present the comparison of the average values of *Maespan*, *Mean_Flowtime* and failure rate $Fail_r$ parameter (see Sec. 5.5.2 in Chapter 5).

It can be observed that both hybrid strategies outperform the $GA - SS - ANN$ algorithm in all but 3 cases. It confirms that game-based models are better adapted for the management of all security requirements in the grid system when compared to standard scheduling models, even if the res

Although the security requirements would imply some additional cost to the users of the grid system, it is worth assuming this cost in order to allocate tasks to trustful resources.

6.5.1 Computational Economy and Game-Based Models

The experimental analysis presented in the previous section show that hybrid GA-based schedulers can be effective in solving the users games, however the main drawback of using such methods is their high computation complexity. The game scenarios presented in Sec. 6.3 are very general, which makes them useful in supporting the users decision process in various situations. In some real-life approaches the game scenarios are usually based on the well-known economical models.

Market-based approaches in grid computing enable grid resource owners, acting as sellers, to earn revenue by allowing others (mainly grid End-users, acting as buyers) to use their (idle) computational resources. The pricing of resources is driven by supply and demand. These models can be easily translated into the game-theoretical frameworks and are useful in grid resource management, as well as in defining users' decision strategies.

Table 6.3 Average values of *Makespan* for $GA-SS-ANN$, $SGA-GA$ and $SGA-PMCT$ algorithms [$\pm s.d.$], ($s.d.$ = standard deviation)

Strategy	Small	Medium	Large	Very Large
		Static Instances		
GA-SS-ANN	4208842.037 [± 210505.265]	4216980.163 [± 249225.887]	4309539.605 [± 263233.057]	4399950.825 [± 290453.201]
SGA-GA	**4104953.259** [± **379579.997**]	**4156536.877** [± **319105.946**]	4264926.597 [± 548415.652]	**4353604.208** [± **595472.951**]
SGA-PMCT	4185298.477 [± 574689.195]	4162755.537 [± 444243.979]	**4260258.291** [± **676018.949**]	4365824.522 [± 487088.573]
		Dynamic Instances		
GA-SS-ANN	4141538.885 [± 247988.145]	4212439.475 [± 342459.080]	4232327.490 [± 333199.727]	4364692.950 [± 339043.674]
SGA-GA	**4064399.586** [± **295082.511**]	**4159942.678** [± **573609.898**]	**4181202.361** [± **503195.319**]	4330472.697 [± 485326.735]
SGA-PMCT	4104009.894 [± 230614.767]	4194715.259 [± 525510.365]	4229959.495 [± 410287.752]	**4329168.378** [± **454255.168**]

The following paragraphs present a general characteristics of the most popular economically- and game-based approaches for modelling users' relations and decisions in scheduling process.

Commodity Market Model

This model is based on the Meta-broker architecture (described in Sec. 1.2.2). It is assumed here that the service providers primarily charge the end user for the resources they consume and the pricing policies are based on the demand from the users and the supply of resources. The resource owners and service providers are selfish in this approach and the end-users may or may not cooperate [26].

Table 6.4 Average values of $Mean\_Flowtime$ for $GA-SS-ANN$, $SGA-GA$ and $SGA-PMCT$ algorithms [$\pm s.d.$], ($s.d.$ = standard deviation)

Strategy	Small	Medium	Large	Very Large
		Static Instances		
GA-SS-ANN	1098725220.445 [± 148984029.042]	2261958805.835 [± 296213971.853]	4395864089.470 [± 403819795.484]	**8705728350.062** [± 779128466.164]
SGA-GA	1080025209.170 [± 106385883.899]	2212272645.989 [± 225632106.035]	7649954581.921 [± 564374456.205]	8790927826,710 [± 820203622.476]
SGA-PMCT	**1039256248.489** [± **132828810.736**]	**2177583973.023** [± **279653260.144**]	**4251057955.321** [± **390873259.314**]	8752787592.196 [±849851282.277]
		Dynamic Instances		
GA-SS-ANN	1163342728.245 [± 136548966.434]	2161846250.347 [± 272493690.708]	4322245472.632 [± 533180226.552]	8734534678.245 [± 835468708.749]
SGA-GA	**1124621170.786** [± **247484952.990**]	**2137984828.270** [± **143846367.588**]	**4254686510.766** [± **407990354.145**]	8720343632.821 [± 931632311.801]
SGA-PMCT	1155781873.098 [± 109523418.319]	2200754844.400 [± 203859979.408]	4293599602.333 [± 350634474.868]	**8712482470.785** [± **912872163.393**]

Auctions

In this model there are two groups of participants: sellers (resource owners) and buyers (grid end-users). The cooperation between users to form a coalition and win the auction is possible, but usually the users behave selfishly. The auction mechanism can be defined in many ways (e.g. English, Dutch, First and Second Price auctions). All of which differ in terms of whether they are performed as open or closed auctions and the offer price for the highest bidder. The users' strategies in particular auctions are discussed e.g. in [52].

Bi-level Synchronized Auctions

The First Price bidding auction mechanism has been extended by Kwok et al. [92] to define the resource management and global scheduling policy at the intra- and inter-site levels in the 3-levels hierarchical grid structure. In the intra-site bidding each machine owner in the site, who acts selfishly, declares the "execution capabil-

Table 6.5 Average values of failure rate $Fail_r$ parameter for $GA - SS - ANN$, $SGA - GA$ and $SGA - PMCT$ algorithms [$\pm s.d.$], ($s.d.$ = standard deviation)

Strategy	Small	Medium	Large	Very Large
		Static Instances		
GA-SS-ANN	3.993% [± 0.98]	4.089% [± 1.56]	8.436% [± 1.67]	8.736% [± 2.09]
SGA-GA	3.877% [± 0.98]	4.356% [± 1.26]	7.543% [± 1.89]	9.015% [± 2.23]
SGA-PMCT	**3.738%** [± **0.92**]	**4.005%** [± 1.05]	7.456% [± **1.35**]	9.832% [± 1.56]
		Dynamic Instances		
GA-SS-ANN	4.880% [± 0.98]	6.097% [± 1.62]	7.456% [± 1.32]	**7.026%** [± **2.11**]
SGA-GA	4.423% [± 0.73]	5.533% [± 0.69]	6.944% [± 0.98]	7.046% [± 1.44]
SGA-PMCT	**3.875%** [± **0.88**]	**4.542%** [± **1.03**]	**5.953%** [± **1.21**]	7.211% [± 1.95]

ity" of the resource. The local manager monitors these amounts and sends a single value to the global scheduler. In the inter-site bidding the global scheduler should allocate tasks according to the values sent by the local dispatchers. The authors prove that the cooperation of the players at both levels are the optimal strategies for both level-auctions. However, for the successful execution of all strategies some synchronization mechanism must be introduced, which can make the system in whole inefficient in a large-scale dynamic environment.

Bargaining Models

In this model the resource brokers bargain with resource providers for lower access price and longer usage duration. The negotiation process is guided by the end-users requirements (e.g., deadline) and can be provided directly between buyers (End-users) and sellers (resource owners). The most recent study on the bargaining cooperative model application in optimizing the energy consumption in grid is proposed in [141].

6.6 Conclusions

This chapter showed the game-theoretic models as the effective methodologies for supporting the grid users' decisions, where the different scheduling criteria, including security and resource reliability, must be considered at the same time. The users' behavior can be effectively translated into the computational model linked to the grid scheduling. Due to large scale of the grid, the non-cooperative games seems to be a potential model for integrating various requirements in grid scheduling.

The users decisions in the scheduling process are modelled by using the two general non-cooperative game scenarios, namely symmetric non-zero sum game and Stackelberg game. The hierarchical procedure of solving those games is complex because of the need of integration and synchronization of two cooperating modules. However, the experimental analysis shows the high efficiency using the meta-heuristics as the resolution methods for game-based models, especially in the case of additional security casts paid by the users. The game-based model concepts can be successfully implemented also in cloud computing, where the secure scheduling and information management remain challenging research problems.

Part IV
Genetic Solutions to Green Scheduling in Computational Grids

Chapter 7
Evolutionary Inspired Solutions for Energy Management in Green Computing: State-of-the-Arts

Abstract. The quest for more powerful computational resources has enabled significant scientific discoveries and also has immensely improved people's daily life. However, such advancement has significantly strained the electrical energy resources, distribution, and protection systems. Therefore, in the past several years, engineers, researchers, and vendors have teamed up to design, develop, and test devices, procedures, methodologies, and algorithms that constrict the use of electrical energy in computing devices. This chapter surveys the field from the perspective of evolutionary inspired solutions for energy management in "green" computing. This survey bridges two distinct research fields: (a) evolutionary computing and (b) green computing. The presented models are classified according to a general taxonomy of energy and resource management methods in large-scale heterogeneous computing systems.

7.1 Taxonomy of Energy Management in Modern Distributed Computing Systems

A significant volume of research has been done in the domain of energy aware resource management in today's large-scale computing system. Following a taxonomy for cloud computing proposed in [51] the management methods in modern distributed computing systems can be classified into two main categories: Static Energy Management (SEM) and Dynamic Energy Management (DEM), as it is shown in Fig. 7.1.

SEM class contains all technologies that are applied for the system components, architecture and software optimization. At the hardware level the system devices can be replaced by the low-power battery machines or nano-processors and the system workload can be effectively distributed. It allows to optimize the energy utilized in the system, storage and data transfer by reducing the number of idle devices and idle periods of active processors. It is important to carefully consider the implementation of software that is executed in the system in order to achieve a high and fast reduction in the energy usage. Even with perfectly designed hardware, poor

J. Kołodziej: Evolutionary Hierarchical Multi-Criteria Metaheuristics, SCI 419, pp. 139–153.
springerlink.com

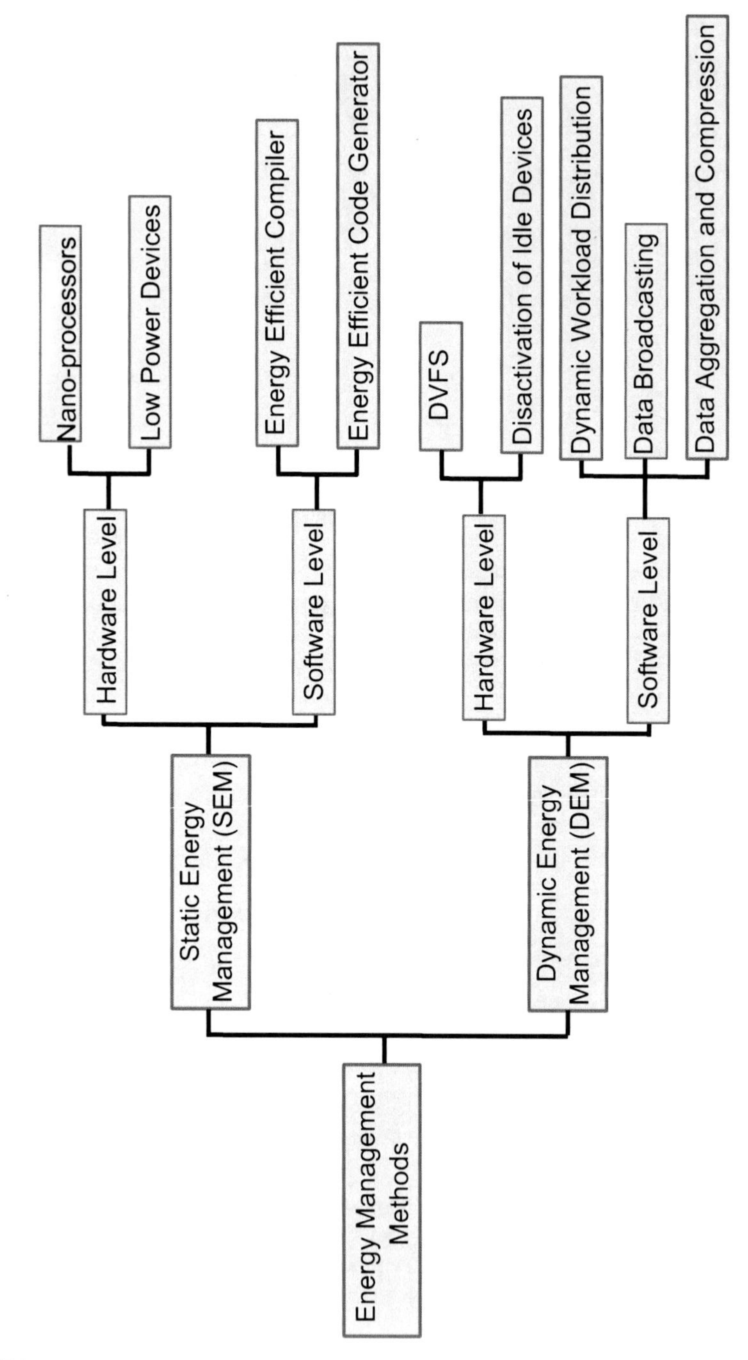

Fig. 7.1 Taxonomy of Energy Management in Large-scale Distributed Computing Systems

software design can lead to significant power and energy losses. Therefore the process of compilation or code generation and the order of instructions in application source code may have an impact on energy utilization.

DEM techniques include the strategies for dynamic adaptation of the system performance to the current states of the resources and system's services. DEM methods enable a dynamic adjustment of power states according to current system's performance. Similarly to the static methods, dynamic management methodologies can be categorizes into hardware– and software–based techniques according to the application levels criterion. Hardware tools can be classified as Dynamic Performance Scaling (DPS), such as Dynamic Voltage and Frequency Scaling (DVFS), and partial or complete dynamic deactivation of idle processors. The software techniques class includes all optimization techniques related to dynamic workload distribution, efficient data broadcasting, data aggregation and dynamic data (and memory) compression.

A fast development of global communication technologies enables an unlimited access of the computing systems' users to all system resources. However, the resource management in such systems remains challenging, because of the different local policies at the system and operational levels and high parametrization and dynamics of the whole structure. Evolutionary and genetic techniques, due to their robustness and abilities of easy hybridization with other approaches, are promising candidate solutions for the resource management in today's high performance systems. However, the class of energy-aware genetic-based optimization methods is not so large. Basically, conventional single-population genetic and evolutionary algorithms are applied to the energy optimization. If energy consumed by the system is a component of a multi-objective fitness function, the Multi-objective Genetic Algorithm (MOGA) framework [46] is a key solution to tackle the complexity of the optimization process. Ant Colony Optimization (ACO) [38] and Particle Swarm Optimization (PSO) [71] algorithms are useful in creating optimal path and tree structures in graph-based models of networks, multi-processor machines and parallel applications. Finally, just few approaches in Grid and Cloud scheduling show that Island and Cellular Parallel GAs (CPGA) can generate the low-cost schedules in various 'energetic' scenarios.

This chapter classifies the recent and most promising evolutionary inspired solutions to the static and dynamic energy-aware resource management in modern large-scale distributed computing systems according to the following four attributes:

- evolutionary search technique (algorithm's type);
- representation of problem solution (chromosome encoding method and structure);
- objective function;
- application environment.

7.2 Evo-Driven Static Energy Optimizers in Embedded Systems

Static methods of energy conservation in distributed systems are mainly employed at the hardware system's level. Energy absorbing devices can be replaced by the machines with low–power batteries and nano-processors. In such a case the meta-heuristics are applied for the optimization of the system's architecture, and optimization of compilers and the generators of the source codes at the software level.

Table 7.1 present three basic genetic-based approaches to static energy management in embedded systems.

Table 7.1 Genetic-based methods for static resource management in embedded systems

Project	Algorithm Type	Chromosome	Objective	System
Energy-aware code generator [98]	single population genetic algorithm	sequence of basic operations with their parameters	a consumed power or energy of a program	embedded systems
Energy-aware code optimizer [9]	single population multi-objective genetic algorithm	sequence of basic operations with their parameters	cycles per instruction and energy dissipation	embedded systems
NSGA for redundancy allocation [103]	Non-dominated Sorting GA (NSGA)	vector of redundancy levels for the resources	system reliability and energy consumption	embedded systems

Most of the embedded systems are composed of the Digital Signal Processors (DSPs) that flexibly account for the modification of the system specification. However, many embedded applications are still prepared in assembly code, that usually leads to incorrect codes. This kind of implementation is time-consuming and inefficient in the utilization of the system energy. For such reasons, there is a need for optimizing compilers and and application source codes to make them capable of exploiting the irregular architecture features of DSPs.

Lorenz et al. define in [98] an energy aware code generator (GCG) based on single population genetic algorithm. This code generator reduces the energy consumption by the optimal selection and scheduling of the instructions in the source codes. The genetic algorithm module works on specialized chromosomes, that encode a set of basic blocks, which are created by using a simple decomposition procedure to the source program. Each such a block is represented by a node in Data Flow Graph (DFG). Each gene of the chromosome represents an information about used registers, performed processor instruction, execution cycle (etc.), which are needed for compilation simple software instructions like 'a load' or 'an addition'. An objective function is defined as an amount of power or energy consumed during a compilation of a code. The authors used their method for 'Single Instruction Multiple Data' (SIMD) instructions. The obtained results show the 30% of the energy reduction with 8% of reduction of the application code.

Similar GA-inspired approach to code optimization is proposed by Azzemi in [9]. The author considers the multimedia DSP processors and defines an architecture-based parametric optimization of C source codes for an iterative compilation. Successive source-level, code transformations are applied in order to evaluate an application expression profile. The optimization criteria are defined as a bi-objective fitness with cycles per instruction and energy dissipation as the components. This function is optimized by a simple MOGA technique. The achieved energy reduction is in the range of 17%.

Optimal utilization and reliability of resources associated to the data consolidation are the key quality attributes in several types of today's complex embedded systems. The energy consumption may be reduced by replication of computational and data nodes. This problem is referred to as *redundancy allocation in embedded systems* . In [103] Meedeniya et al. try to solve the redundancy allocation problem in the embedded systems by using the Markov Reward Model [70] for system representation. The authors optimized a bi-objective function with system reliability and energy consumption components by using the Non-dominated Sorting GA (NSGA) . Each chromosome encodes a single redundancy allocation. Each allele in a chromosome represents a redundancy level for a system component. The experimental results show that the proposed method can significantly reduce the energy consumption for a very small trade-off of reliability, which would definitely be an interesting information for the system designer.

7.3 Evolutionary Inspired Dynamic Data and Resource Management in Green Computing

Evolutionary-based solutions to dynamic energy management in large-scale distributed systems are primarily proposed as scalable and robust methodologies for scheduling and data processing in networking, cluster and grid computing. This section highlights the most recent research in energy aware grid and cloud scheduling, where the voltage supply of the devices may be modulated in the system. The genetic support in data broadcasting and aggregation in wireless sensor networks is also discussed in this section.

7.3.1 Dynamic Voltage and Frequency Scaling in Energy-Aware Resource Allocation and Scheduling in Distributed Computing Systems

Scheduling in conventional distributed systems has been mainly studied for system performance parameters without data transmission requirements. With the emergence of Data Grids (DGs) and data centers, data-aware scheduling has become a major research issue. Today's data centers arise quite naturally to support needs of scientific communities to share, access, process, and manage large data collections.

Computing devices (CPUs) are the major energy "consumers" in a data center. The energy of the system is utilized for the tasks execution, data storage at the data hosts, data transmission, decoupling of data from processing and data replication.

Power and total energy consumption can be reduced by lowering the supply voltage of CPUs by using the *Dynamic Voltage Scaling (DVS)* or *Dynamic Voltage and Frequency Scaling (DVFS)* methods [97]. It is assumed in these models that each machine in the system (it can be a data or computing node) is equipped with a DVS module. It allows to modulate the supply voltage and operating frequencies of the resources. Instead of complete deactivation of the processors, the clock frequencies along with adjustments of the supply voltage can be gradually reduced or increased in cases when the resources are not fully utilized.

The energy consumption model in the data center is usually based on the power consumption model in Complementary Metal-Oxide Semiconductor (CMOS) logic circuits. The power consumption of a CMOS-based microprocessor is defined as a sum of the capacitive, short-circuit and leakage powers. The most significant factor is the capacitive power , which can be interpreted as the dynamic power consumption Pow_d of a CPU at the data center and can be calculated in the following way:

$$Pow_d = A \cdot C \cdot v^2 \cdot f, \tag{7.1}$$

where A is the number of switches per clock cycle, C is an effective switched capacitance of the circuits, v is the supply voltage and f is the clock frequency. Assuming that A and C are the constant parameters of CPUs (or machines), it can be observed for Eq. 7.1 that power is proportional to v^2. Energy consumption of the processor can be expressed as processor power multiplied by execution time of the computing application. Therefore, the decrease in voltage supply can reduce the energy per operation in a quadratic manner. Unfortunately, it may also significantly increase the completion time of the computation. The detailed energy model description in grid and cloud data centers can be found in [73] and [51].

The DVFS technique is classified as an effective hardware dynamic energy optimizer in resource allocation and scheduling problems in large-scale distributed systems. The energy-aware scheduling is usually considered as a multi-objective global optimization problem with makespan and cumulative energy consumption as the main criteria. In most of the DVFS approached the scheduling has been defined as classical or dynamic load balancing problem. In such cases linear, dynamic and goal programming are the major optimization techniques (see i.e. [94], [169], [74], [73],

Table 7.2 Selected population-based meta-heuristic for energy aware scheduling with modular voltage supply

Project	**Algorithm Type**	**Chromosome**	**Objective**	**Application Area**
Price Guided GA [132, 133]	single population GA with modified mutation operation (shadow price index for each task machine pair and classical move or swap operation)	vector of tasks labels	total energy consumption	cloud computing
Energy Aware GA Grid Scheduler [82]	combinatorial GA (specified genetic operations) with various replacement techniques	schedule representation for independent batch scheduling in grids	makespan (privileged) and total energy consumption	grid computing
Parallel Hybrid GA-based Scheduler [72]	Parallel Multi-objective GA (PMOGA) – Island based model multistart– hybridized with energy-conscious scheduling heuristics (ECS)	length: number of tasks, each gene is defined by task-machine-voltage triplet	makespan (privileged) and total energy consumption	computing and embedded systems
Parallel Hybrid GA-based Scheduler [105]	Parallel Multi-objective GA (PMOGA)– island based model and multistart– hybridized with energy-conscious scheduling heuristics (ECS)	length: number of tasks, each gene is defined by task-machine-voltage triplet	makespan and total energy consumption (simultaneously)	cloud computing
Multiobjective Hybrid GA Scheduler [106]	MOGA is hybridized with simulated annealing for scheduling problem in chip multiprocessor (CMP) system	schedule representation	total execution time of tasks and the system energy consumption (optimized simultaneously)	cluster computing

[77]). Recent evolutionary-based approaches with DVFS method applied to reducing the energy consumption are presented in Table 7.2. The total energy utilization in the system is a component of the fitness function.

In [132] and [133] Shen et al. present a *Shadow Price GA* technique for improving the genetic operations in standard GA used as a scheduler in computational cloud. The 'shadow price' for a pair task-machine is defined as the average energy consumption per instruction for the processor that can operate at different voltage levels. Then the classical move and swap mutation operations are used for an optimal mapping of tasks to machines. A total energy consumption in the system is defined as a fitness function for shadow price GA scheduler.

Total energy consumed by a computational grid is the key criterion in independent batch scheduling problem discussed by Kołodziej, Khan and Xhafa in [82]. The expected times of the execution of tasks on the machines in the system are estimated by using the *Expected Time to Compute* matrix model [5]. Two implementations of single-population GA-based schedulers were developed for makespan and energy consumption optimization. The authors consider two scenarios, where all machines works at the highest voltage level and are switched to the sleep mode in idle periods, and the case of operating at different voltage levels under optimal makespan constraint. The schedulers were experimentally evaluated in static and dynamic grid environment. In both cases the modulation of the voltage supply of the machines reduced the energy consumption by 25–30 % in average.

Kessaci et al. in [72] present two versions of Multi-objective Parallel Genetic Algorithm (MOPGA) hybridized with Energy-Conscious Scheduling heuristic (ECS). The GA engine is based on the concepts of Island GA and multistart GA models. The authors consider parallel applications represented by a Directed Acyclic Graph (DAG), which are mapped onto multi-processors machines. The voltage and frequencies of the processors are scaled up at 16 discrete levels and genes in GA chromosomes are defined by the task-processor labels and processor voltage. The objective function is composed of two criteria: makespan and cumulative energy consumption in the system. The reduction of the energy utilization achieved in the experimental analysis is about 47.4%. Mezmaz et al. in [105] present an application of the aforementioned methodology in computational cloud. The energy conservation rate in cloud system is very similar to the results presented in [72] .

Another hybrid GA approach is presented by Miao et al. in [106]. The authors propose a multi-objective genetic algorithm hybridized with simulated annealing for solving the scheduling problem a the Chip Multiprocessor (CMP) cluster system. The total execution time of all tasks and the total energy consumption in the system are the main components of the objective function.

7.3.2 Energy Efficient Data Transmission

Data transmission, sometimes referred to as *data broadcasting*, is, together with the resource allocation and scheduling, a fundamental problem in large-scale data centers, intelligent networks and grid and cloud environments. This problem is important in today's large-scale wireless networks such as ad-hoc and sensor networks , where the nodes, acting potentially both as routers and hosts, are equipped with antennas for sending and receiving information. Communication among such nodes may be performed by one-to-one transmissions (single-hop) or by using other nodes as relay stations (multi-hop) . In both cases each sender node must adjust its emission power in order to reach the respective receiver node. Additionally, in the cases where energy is supplied by batteries, the network lifetime is limited by the batteries of the wireless devices. Therefore, energy saving is critical in all network operations.

Minimum Energy Broadcast (MEB) is defined as a problem of minimizing the energy during the data transfer. Formally, it can be formulated as the minimal spanning tree task ($T = (V, E_T)$) in the fully connected graph $G = (V, E)$ representing the system structure. The root of the tree is the source node for the data (signals) emission and the following energy emission function is minimized:

$$P(T) = \sum_{i \in V} \max_{(i,j) \in V_T} d(i,j)^{\alpha}, \tag{7.2}$$

where $d(i,j)$ is the Euclidean distance between the nodes i and j and α is a parameter that, depending on the environment, takes typically values between 2 and 4. It is assumed that the graph G for wireless networks is directed and $d(i,j)^{\alpha} < p_{max}$, where p_{max} is a maximal emission power in the system. If the antennas in the network nodes are directional, a beamwidth and a beam direction must be chosen for each node $i \in V$.

In classical cluster and grid systems the energy utilized for the data files transfer between two connected nodes is summarized (and then optimized) for all possible nodes pairs.

Table 7.3 presents the evolutionary inspired approaches to energy aware data transfer in cluster system and wireless sensor networks.

Hernández, Blum and Francès in [62], address the problem of signal broadcasting in the ad-hoc networks. They consider the system with omni-directional and directional antennas . The emission energy defined in Eq. 7.2 is an objection function, which is globally optimized by using a specialized Ant Colony Optimization algorithm – Min-Max Ant System in the Hyper-Cube Framework [140]. At each iteration of the algorithm artificial ants construct a broadcasting tree rooted at the emission source node. Local search r-shrink algorithm is applied to each of these trees and the pheromone values may be updated by using also the best-so-far solutions. The power saving rates achieved in the experimental analysis is ab. 85 %, which makes the methodology spectacular solution for improving the network effectiveness in the reduction of the energy emission.

Cao et al. in [28] have considered a routing problem in Wireless Sensor Networks (WSNs) , in which a node and its cluster-head engage in a multi-hop communication. They used a PSO algorithm for clustering the network nodes. A distance based minimum spanning tree of the weighted graph of the network is generated and the best connection between a node and its cluster-head is searched from all the optimal spanning trees according to the criterion of energy consumption. Cluster-heads are

Table 7.3 Evo-based solutions for energy aware data broadcasting

Project	Algorithm Type	Chromosome	Objective	System
Ant Colony Optimizer for Broadcasting [62]	Ant Colony Optimization (ACO) algorithm - Min-Max Ant System in the Hyper- Cube Framework	emission spanning tree nodes	total energy emission	ad-hoc networks
MST-PSO for clustering and routing problem [28]	Particle Swarm Optimization (PSO)	vector of nodes of spanning trees in the network cluster	total energy emission	wireless sensor networks
Cellular Genetic Scheduler [60]	cellular GA (cGA) for combinatorial problems	vector of cluster labels for subtasks in the application	makespan (privileged) and total energy consumption	cluster computing system
NSGA-II for data compression [102]	Non-dominated Sorting GA-II (NSGA-II)	binary vector of quantization parameters	information entropy and the number of distinct quantization levels (used in the quantizer	wireless sensor networks

elected based on the energy available to the nodes and the Euclidean distance to its neighbor node in the optimal tree. The results show that the PSO-based clustering methods ensure longer network life compared with the reliability of the system with its original architecture.

An interesting approach of cellular GA-based schedulers to cluster computing is presented by Guzek et al. in [60]. The authors consider a general scheduling problem of parallel application modeled by a DAG in a cluster of heterogeneous machines. The cellular algorithm is used primarily for sub-tasks clustering (the number of clusters is proportional to the number of processors in the machine) and scheduling. The primary objection is makespan and the second criterion – total energy consumed during the inter-processor communication. This communication model is based on the classical delay model [122] and the energy utilized for a data transfer is measured for each CPUs connection in a parallel machine.

Dynamic data compression in the application codes seems to be a promising software tool for saving the energy used for the data propagation in wireless sensor networks. Compression methods exploit the data structure and reduce the data size. Marcelloni and Vecchio [102] perform a data compression on a network (single) node based on a differential pulse code modulation scheme with quantization of the differences between consecutive codes of the signal samples. The trade-off between a performance of compression algorithm and the amount of the lost information is determined by the set of quantization parameters. The authors employ the Non-dominated Sorting Genetic Algorithm II (NSGAII) for optimizing the combinations of these parameters corresponding to different optimal trade-offs. The chromosomes in this approach encode quantizers defined by using the following parameters:

- width of the dead zone;
- width of the cell in the first granular subregion;
- number of cells in the first granular subregion;
- width of the cell in the second granular subregion; and
- number of cells in the second granular subregion.

The chromosomes are represented by binary Gray strings. The granular regions are the regions with quantization levels. Information entropy and the number of distinct quantization levels (used in the quantizer are the optimization criteria. The evaluation analysis of the proposed method shows the 62% reduction of the energy consumed in data transmission.

7.3.3 The Workload Placement Problem – The Data Aggregation

The data aggregation is the combination of data from different sources according to a certain aggregation function, e.g., duplicate suppression, minima, maxima and average. A big amount of the energy in data centers is the idle power wasted when servers run at low utilization. Multiple data center applications may be hosted on a common set of servers. Also sensor nodes in wireless networks may generate significant volume of the redundant data. This allows for consolidation of application workloads on a smaller number of servers and aggregation of similar data packets from multiple network nodes that may increase the system utilization by save the energy.

In grid and Cloud computing the problem of loading servers to a desired utilization level for each resource may be modelled as a multi-dimensional bin packing problem where servers are bins with each resource (CPU, disk, etc.) being one dimension of the bin. The bin size along each dimension is given by the energy optimal utilization level. Each hosted application with known resource utilizations can be treated as an object with a given size in each dimension. The ultimate goal of the consolidation algorithm is to pack all items to possible minimal number of bins. An objective function for such a problem can be defined as follows (see also [44]):

$$f = \sum_{v=0}^{n-1} y_v, \tag{7.3}$$

and is minimized subject to the following constraints:

$$\sum_{i=0}^{m-1} \bar{r}_{i,k} x_{i,v} \leq C_{v,k} y_v, \forall v \in \{0,\ldots,n-1\}, \forall k \in \mathbb{R} \tag{7.4}$$

$$\sum_{v=0}^{n-1} x_{i,v} = 1, \forall i \in \{0,\ldots,m-1\} \tag{7.5}$$

where

- n is the number of bins;
- m is the number of items;
- y_v is the bin variable which is 1 if the bin v is selected and 0 otherwise;
- $x_{i,v}$ is the allocation variable equals 1 if the item i is assigned to the bin v, and 0 otherwise;
- $C_{v,k}$ is the capacity of bin v of resource $k \in \mathbb{R}$;
- $\bar{r}_{i,k}$ is the i-th item maximum demand for resource $k \in \mathbb{R}$ over the last measurement period.

The condition 7.4 ensures that the capacity of each bin is not exceeded and constraint 7.5 guarantees that each item is assigned to at most one bin.

In wireless sensor networks signal processing methods may be used for data aggregation. In this case, it is referred to as data fusion where a node is capable of producing a more accurate output signal by using some techniques such as beam forming to combine the incoming signals and reducing the noise in those signals.

Selected genetic-based methods for data aggregation are reported in Table 7.4.

One of the most recent ACO approaches in data aggregation in cloud computing is presented by Feller et al. in [44]. The authors used ant colony optimization technique for the consolidation of virtual machines on the least number of physical

Table 7.4 Genetic-based methods for data aggregation in large-scale distributed systems

Project	Algorithm Type	Chromosome	Objective	System
Energy-aware ACO data aggregator in clouds [44]	Ant Colony Optimization (ACO) algorithm	binary vector of bin variables	the number of active servers in the system and energy utilization	cloud computing
ABC-PSO [146]	Particle Swarm Optimization (PSO) algorithm hybridized with ant-based control technique	vector of sensor nodes	PSO algorithm determines a local thresholds and decision error is minimized	wireless sensor networks

nodes in the cloud system. The problem is interpreted as an instance of the Multi-Dimensional Bin-Packing (MDBP) problem . The fitness function is defined as the sum of boolean bin variables given by the Eq. 7.3. The authors follow the MAX-MIN Ant System (MMAS) Framework for updating the pheromone trials [140] for the ants. In order to estimate the energy consumed by a workload placement, the authors approximate the power function of a host, which is defined as a linear function $P(u)$ of the host utilization u, i.e.:

$$P(u) = (P_{max}, \dots, PPidle) \times u + P_{idle}, \tag{7.6}$$

where P_{idle} and P_{max} denote the average power values when the system is idle and fully utilized, respectively. Computational results show the 4.1% of energy conservation on average 4.7% of hosts.

High effectiveness in the data-aggregation in wireless sensor networks can be achieved by the determination of optimal local thresholds in the decisions made for detecting the events. Each sensor node in the network collects local observations corrupted by noise and sends a summary to a fusion center, which is responsible for making the final decision. Thresholding may lead to a gain in terms of bandwidth and energy consumed by the system. Veeramachaneni et al. [146] present a hybrid of ant-based control and PSO (ABC-PSO) method for the local threshold management to achieve an optimal decision route. Partial solutions to the optimization problem are constructed by artificial ants that move from a node to another and define the paths of network nodes. Then PSO algorithm identifies the thresholds and achieves the minimum error for the sequence. The feedback on this is presented to ants to help them to improve the qualities of node sequences to achieve optimal thresholds on all nodes and an optimal decision route (hierarchy) that assure minimum energy expenditure.

7.4 Conclusions

This chapter surveyed the recent research results related to the evolutionary inspired methodologies supporting the energy and power management in modern large-scale distributed computing systems. Although, the genetic meta-heuristics are still not the most popular solutions to the green computing problems, the experimental results briefly analyzed in this chapter, confirm the efficiency of the genetic techniques in the reduction (in the range of 6 to 85%) of energy consumed for computing.

Table 7.5 presents the summary of the evolutionary approaches to energy management in today's distributed computing systems, as a critical analysis of state-of-the-art in 'evolutionary-driven green computing'.

The algorithms in Table 7.5 are classified into mono- and multi-population categories. The first group contain conventional single population genetic techniques for global optimization. They are employed as promising tools for solving all problems addressed in this chapter and work in all types of dynamic environments discussed in this survey. The second class of multi-population algorithms is very small, that confirms an early stage of the research in this field.

Table 7.5 Summative analysis of evolutionary-based approaches to energy management in distributed computing systems

Algorithm Class	Algorithm Type	Application Areas	Problems to Which Applicable
Mono-population	Basic GA	grid and cloud computing, embedded systems	scheduling and resource allocation, code optimization
	ACO	ad-hoc networks, cloud computing	data broadcasting, data aggregation
	PSO	wireless sensor networks	data broadcasting,data aggregation
	NSGA	embedded systems	data aggregation
	NSGA-II	wireless sensor networks	data compression, source code optimization
	MOGA	cluster computing, embedded systems	scheduling and resource allocation, code optimization
Multi-population	Cellular GA	cluster computing	data broadcasting
	Island and Multi-start Parallel GAs	cluster, grid and cloud computing, embedded systems	scheduling and resource allocation

An emergence of new generation IT systems, implies new challenges in efficient management of huge packages of highly parameterized data. A promising research direction, which can make a significant progress in green computing, is the utilization of game-theoretical models and evolutionary-based resolution methods for supporting the decisions of the system's users and resource providers. Simple cooperative games with Nash-bargained solutions have been already developed and successfully applied in energy-aware scheduling in grids and data centers (see [76], [141]), which can be a basis for evolutionary inspired solvers of such models. However, these models may not reflect the dynamic nature of the large-scale computational systems. An incorporation of new additional criteria into the energy-aware data and resource management in future generation distributed systems may expose the low efficiency of existing solutions at both hardware and software levels. It certainly implies a need of development of new models and meta-heuristic optimization techniques which can tackle the demand of the system components, new access polices and conditions for the resources, and users' special preferences and requirements.

Chapter 8
Energy-Aware Scheduling of Independent Tasks in Computational Grids

Abstract. This chapter introduces the application of the Hierarchical Genetic Strategy-based Grid scheduler (HGS-Sched) to the energy-aware independent batch scheduling problem in Computational Grids (CGs). The *Dynamic Voltage Scaling (DVS)* methodology is used for both scaling the power supply of the grid resources and reducing the cumulative power energy utilized by the grid computing machines. Two implementations of HGS-Sched—with elitist and struggle replacement mechanisms respectively—are defined and empirically evaluated. The effectiveness of the hierarchical schedulers are compared with the quality of single-population Genetic Algorithms (GAs) and Island GA models for four CG significant scenarios in static and dynamic modes. The simulation results show that meta-heuristic grid schedulers can significantly reduce the energy consumption in the system as well as be easily adapted to various scheduling scenarios.

8.1 Introduction

The main issues related to power consumption and effective thermal management in high performance computing have been induced by the sheer scale of enterprise computing environments and data centers. In large supercomputer centers and next generation distributed systems such as 'green' grid clusters and clouds, the growing operating, power and cooling rates have become the dominant part of the users' and system managers' budgets. Novel innovative green computing technologies are mainly devoted to the optimization of system thermodynamics [108]. Profiles of hardware energy consumption and application energy consumption are gathered in order to be correlated with workload distribution and energy consumption of the different power and cooling systems [43].

While CGs have been widely promoted as affordable alternatives to supercomputers, a significant disproportion of resource availability and resource provisioning has been empirically observed [76]. Therefore, a significant deal of research in grid computing is devoted to design novel, effective grid schedulers, which can simultaneously optimize the key grid objectives—such as makespan, flowtime, and

J. Kołodziej: Evolutionary Hierarchical Multi-Criteria Metaheuristics, SCI 419, pp. 155–175.
springerlink.com

resource utilization [35], as well as the energy consumed by all system components and users.

That is to say, while the main purpose of grid schedulers is to efficiently and optimally allocate application tasks to a set of available resources, one should also consider a series of requirements including energy efficiency. Energy-efficient scheduling in CGs has therefore become a relevant yet complex endeavor due to the multitude of constraints and the different optimization criteria and priorities of the resource owners. Heuristic approaches have demonstrated to be effective for designing energy-aware grid schedulers by keeping a balance among various preferences and goals of the grid users, resource and service managers, and resource owners.

This chapter addresses the problem of energy optimization for *Independent Batch Scheduling* in CGs. The *average energy consumption* is considered as a complementary scheduling criterion along with the *makespan* as the primary objective. According to the notation introduced in Sec. 1.4.2, an instance of the independent batch grid scheduling problem with energy optimization criterion is expressed in the following way:

$$Rm\left[\{b, indep, (stat, dyn), hier\}\right](C_{max}, E_I(E_{II})) \tag{8.1}$$

where:

- C_{max} – denotes a makespan as the primary scheduling objective
- $E_I(E_{II})$ – denotes total energy consumption as the second scheduling criterion (E_I or E_{II} is selected depending on the scheduling scenario (see Sec. 8.3.2))

This chapter extends the empirical analysis presented in [82] by the implementation and the comparative analysis of the effectiveness of the multi-population and single-population GA-based Grid schedulers. The term 'green' used in this chapter refers not just to the low-power system devices, but also to the energy-aware schedulers.

8.2 Energy Model

The main module in the energy-aware grid scheduling model presented in this chapter is the *Dynamic Voltage and Frequency Scaling (DVFS)* technique. Used for adjusting the voltage supplies and frequencies of the grid computational nodes, the DVFS technique is primarily based on the power consumption model employed in complementary metal-oxide semiconductor (CMOS) logic circuits [13]. In this model, the capacitive power Pow_{ji}–utilized by the machine i for computing the task j–depends on the voltage supply and machine frequency, and it is calculated as follows:

$$Pow_{ji} = A \cdot C \cdot v^2 \cdot f, \tag{8.2}$$

where A is the number of switches per clock cycle, C is the total capacitance load, v is the supply voltage and f is the frequency of the machine (see also Chapter 7, Sec. 7.3.1, Eq. (7.1)). The energy consumed per machine i for the computation of task j can then be derived using the following formula:

$$E_{ji} = \int_0^{completion[j][i]} Pow_{ji}(t)dt \tag{8.3}$$

where $completion[j][i]$ is a completion time of the task j on machine i.

Each machine in the grid has been equipped with a DVFS module [97] for scaling its supply voltage and operating frequency. It has been assumed that the frequency of the machine is proportional to its processing speed (see [104]). It follows from the Eq. (8.2) that the reduction of the supply voltage and frequency is directly correlated to the reduction of the energy utilization. Table 8.1 shows the parameters for 16 DVFS levels and three main 'energetic' categories for machines defined for the grid system employed in this study.

Table 8.1 DVFS levels for three machine classes

	Class I		**Class II**		**Class III**	
Level	**Volt.**	**Rel.Freq.**	**Volt.**	**Rel.Freq.**	**Volt.**	**Rel.Freq.**
0	1.5	1.0	2.2	1.0	1.75	1.0
1	1.4	0.9	1.9	0.85	1.4	0.8
2	1.3	0.8	1.6	0.65	1.2	0.6
3	1.2	0.7	1.3	0.50	1.9	0.4
4	1.1	0.6	1.0	0.35		
5	1.0	0.5				
6	0.9	0.4				

The energetic class of machine i, $(i \in M)$, denoted by s^i and represented by the meta-vector $Vr_{(i)}$ of DVFS levels, can be specified as:

$$Vr_{(i)} = \left[(v_{s_0}(i), f_{s_0}(i)); \ldots ; (v_{s_{l(max)}}(i), f_{s_{l(max)}}(i))\right]^T \tag{8.4}$$

where $v_{s_l}(i)$ refers to the voltage supply for machine i at level s_l, $f_{s_l}(i)$ is a scaling parameter for the frequency of the machine at the same level s_l, and l_{max} is the number of levels in the class s^i. The parameters $\{f_{s_0}(i), \ldots, f_{s_{l(max)}}(i)\}$ are in the $[0,1]$ range and should be interpreted as the relative frequencies of the machine i of class s^i at the $s_0, \ldots, s_{l(max)}$ DVFS levels.

The reduction of the machine frequency and its supply voltage can lead to the extension of the computational times of the tasks executed on that machine. For a given 'task-machine' pair (j,i), the completion times for the task j on machine i at different DVFS levels in the class s^i can be interpreted as the coordinates of a vector $\widehat{ETC}[j][i]$ which is defined as:

$$\widehat{ETC}[j][i] = \left[\frac{1}{f_{s_0}(i)} \cdot ETC[j][i], \ldots, \frac{1}{f_{s_{l(max)}}(i)} \cdot ETC[j][i]\right] \tag{8.5}$$

where $ETC[j][i]$ are the expected completion times for task j on machine i calculated by using the conventional ETC matrix model (see Chapter 2, Sec. 2.2).

The ETC matrix model defined in Chapter 2 can be directly adapted to the energy-aware scheduling of independent tasks in grids. The completion times calculated for each pair (j,i) of task–machine labels in conventional ETC matrix (see Eq. (2.2)) should be replaced by the $\widehat{ETC}[j][i]$ vectors, that is:

$$\widehat{ETC} = \left[\widehat{ETC}[j][i][s_l]\right]_{n \times m \times s_{l(max)}} \tag{8.6}$$

where $\widehat{ETC}[j][k][s_l]$ is the time necessary for the completion of the task j on machine i at the level s_l.

Based on Equations (8.2), (8.3) and (8.6) the energy utilized for completing task j on machine i at level s_l can be defined as a scalar product of the number of switches per clock cycle, the total capacitance load, the frequency and the squared voltage at level s_l, and the estimated completion time of task j on machine i. That is to say:

$$E_{ji}(s_l) = \gamma \cdot (f_{s_l}(i))_j \cdot f \cdot [(v_{s_l}(i))_j]^2 \cdot \widehat{ETC}[j][i][s_l], \tag{8.7}$$

where $\gamma = A \cdot C$ is a constant parameter for a given machine class, $(v_{s_l}(i))_j$ is a voltage supply value for class s^i and machine i at level s_l for computing task j, and $(f_{s_l}(i))_j$ is a corresponding relative frequency for machine i.

Based on the Equations (8.6), (8.7) and (8.5) the computational times for each possible pair (j,i) at the level s_l can be calculated as:

$$\begin{aligned} Tim_{\{j,i,s_l\}} &= \gamma \cdot (f_{s_l}(i))_j \cdot f \cdot [(v_{s_l}(i))_j]^2 \cdot (f_{s_1}(i))_j \cdot ETC[j][i] = \\ &= \gamma \cdot f \cdot [(v_{s_l}(i))_j]^2 \cdot ETC[j][i] \end{aligned} \tag{8.8}$$

The cumulative energy utilized by the machine i for the completion of all tasks from the batch that are assigned to this machine, is defined in the following way:

$$\begin{aligned} E_i = &\sum_{\substack{j \in Tasks(i) \\ l \in \hat{L}_j}} \{Tim_{\{j,i,s_l\}}\} + \gamma \cdot f \cdot [v_{s_{max}}]^2 \cdot ready_i + \gamma \cdot f_{s_{min}}(i) \cdot f \cdot \\ &\cdot [v_{s_{min}}(i)]^2 \cdot Idle[i] = \gamma \cdot f \cdot \sum_{\substack{j \in Tasks(i) \\ l \in \hat{L}_i}} ([(v_{s_l}(i))_j]^2 \cdot ETC[j][i]) + \\ &+ [v_{s_{max}}(i)]^2 \cdot ready_i + f_{s_{min}}(i) \cdot [v_{s_{min}}(i)]^2 \cdot Idle[i] \end{aligned} \tag{8.9}$$

where $Tasks(i)$ is a set of tasks assigned to machine i, $ready_i$ is the ready time of machine i, $Idle[i]$ denotes an idle time of machine i, and $\hat{L}_i$ denotes a subset of DVFS levels used for the tasks assigned to machine i. All additional machine frequency transition overheads are ignored. These overheads take usually a negligible amount of time (e.g., 10ms- 150ms, see [109]) and do not bear down on the overall ETC model with an active 'energetic' module.

Finally, an average cumulative energy utilized by the grid system for completion of all tasks in the batch is defined as follows:

$$E_{batch} = \frac{\sum_{i=1}^{m} E_i}{m} \tag{8.10}$$

This model is used in the following section for specification of two scheduling scenarios and the definition of the scheduling criteria.

8.3 Scheduling Scenarios and Objectives

Two main scheduling scenarios are considered in this study, namely:

I. **I – Max-Min Mode**, in which each machine works at the **maximal** DVFS level during the execution and computation of tasks and enters into idle mode after the execution of all tasks assigned to this machine;
II. **II – Modular Power Supply Mode**, in which each machine can work at **different** DVFS levels during the task executions and can then enter into idle mode.

In the former, the consumption of the of the energy depends on the 'energetic' class of the system devices or services, defined as 'machines' (resources) in the system. No modifications of the conventional scheduling procedures and standard scheduling objectives—such as makespan, flowtime, tardiness, etc. (see Chapters 1 and 2)—are needed. In the latter, the optimal power supply levels can be specified for each machine, and the energy consumption can subsequently be reduced by diminishing the power supply in the machines while preserving the deadline constraints for the main tasks.

The procedures for calculation and optimization of the two scheduling objective functions, makespan and cumulative energy utilized by the system, are different in the aforementioned scheduling scenarios. The details are discussed in the two following subsections.

8.3.1 Makespan Optimization

The minimization of the makespan is the first step of the optimization procedure in the scheduling objectives. Based on the ETC matrix model and denoted by C_{max}, the makespan can be defined in terms of the completion times of the machines (see Chapter 2, Sec. 2.2.2). The finishing time for the last task in the batch is specified as the maximal completion time of all machines available in that batch. Denoted by $completion[i]$, the completion time of machine i is the cumulative time required for both reloading the machine i after finalizing the previously assigned tasks and for completing the tasks currently assigned to the machine.

In **Max-Min Mode** such completion time can be defined as:

$$completion_I[i] = ready_i + \sum_{j \in Tasks(i)} ETC[j][i]. \tag{8.11}$$

The makespan in this mode is calculated in the following way:

$$(C_{max})_I = \max_{i=1}^{m} completion_I[i]. \tag{8.12}$$

The idle time for machine i working in **Max-Min Mode** can be expressed as the difference between the makespan and $completion_I[i]$, i.e.:

$$Idle_I[i] = (C_{max})_I - completion_I[i] \tag{8.13}$$

It should be clear that for the machine with the maximal completion time (makespan) the idle factor is zero.

In **Modular Power Supply Mode**, for each task-machine pair, the DSV level s_l must be specified. The formulae for computing the completion time, makespan, and idle time at the level s^i can be defined as:

$$completion_{II}[i] = ready_i + \sum_{j \in Tasks(i)} \frac{1}{f_{s_l}(i)} \cdot ETC[j][i]. \tag{8.14}$$

$$(C_{max})_{II} = \max_{i=1}^{m} completion_{II}[i]. \tag{8.15}$$

$$Idle_{II}[i] = (C_{max})_{II} - completion_{II}[i] \tag{8.16}$$

8.3.2 *Energy Optimization*

The second step of the scheduling optimization procedure is the minimization of the total energy consumed in CG for scheduling a given batch of tasks.

The average energy consumed in the system in **Min-Max Mode** is defined as:

$$\begin{aligned} E_I = \tfrac{1}{m} \cdot \textstyle\sum_{i=1}^{m} \gamma \cdot completion_I[i] \cdot f \cdot [v_{s_{max}}(i)]^2 + \\ + \tfrac{1}{m} \cdot \textstyle\sum_{i=1}^{m} \gamma \cdot f_{s_{min}}(i) \cdot [v_{s_{min}}(i)]^2 \cdot Idle_I[i] \end{aligned} \tag{8.17}$$

In **Modular Power Supply Mode** the average cumulative energy is given by Eq. (8.10):

$$E_{II} = E_{batch} = \frac{\sum_{i=1}^{m} E_i}{m} \tag{8.18}$$

where[1]

$$E_i = \gamma \cdot f \cdot \sum_{\substack{j\in T(i) \\ l\in L_i}}([(v_{s_l}(i))_j]^2 \cdot ETC[j][i]) + \\ + [v_{s_{max}}(i)]^2 \cdot ready_i + f_{s_{min}}(i) \cdot [v_{s_{min}}(i)]^2 \cdot Idle[i] \quad (8.19)$$

In both cases E_I and E_{II} are minimized and subject to the following constraint:

$$\sum_{l\in \hat{L}_i} \left[\frac{1}{f_{s_l}(i)} \cdot ETC[j][i]\right] \leq C_{max};\ \forall i \in \{1,\ldots,m\}, \quad (8.20)$$

where $\hat{L}_i$ denotes a subset of DVFS levels specified for tasks assigned to machine i.

8.4 Empirical Analysis

The implementation of the energy management model based on the DVFS method typically produces an improvement in the load balancing of machines. However, DVFS itself does not change the task assignment. While machines are kept in use for a longer time, they work in a low-cost mode in terms of energy consumption. Conventional load-balancing methods, such as goal programming, are typically effective just for the static scheduling case. It is arguable that, for a successful implementation of dynamic load-balancing and dynamic programming schedulers, a knowledge of all possible states in the system is needed, which is not feasible for large-scale grids.

Metaheuristic approaches are the most promising solution for 'green' scheduling. The effectiveness of single-population GAs for the energy optimization in grid scheduling has been presented in [82]. This chapter substantially extends our initial analysis by introducing an empirical evaluation of multi-population grid schedulers. Within this section, the implementations of the *HGS-Sched* algorithm for energy-aware scheduling problem is referred by *Green-HGS-Sched*.

Similarly to previous chapters, several grid scenarios for static and dynamic scheduling are modeled using the grid simulator. In this case the general architecture of *Sim-G-Batch* has been extended through an energy module as presented in Fig. 8.1.

The simulator generates benchmarks for the problem based on the following input data:

- workload vector of tasks;
- computing capacities of machines;
- prior machines loads;
- machine categories specification parameters (number of classes, maximal computational capacity value, computational capacity ranges interval for each class, machine operational speed parameter for each class, etc.);
- DVFS levels matrix for machine categories; and
- the ETC matrix.

[1] See Equation (8.9).

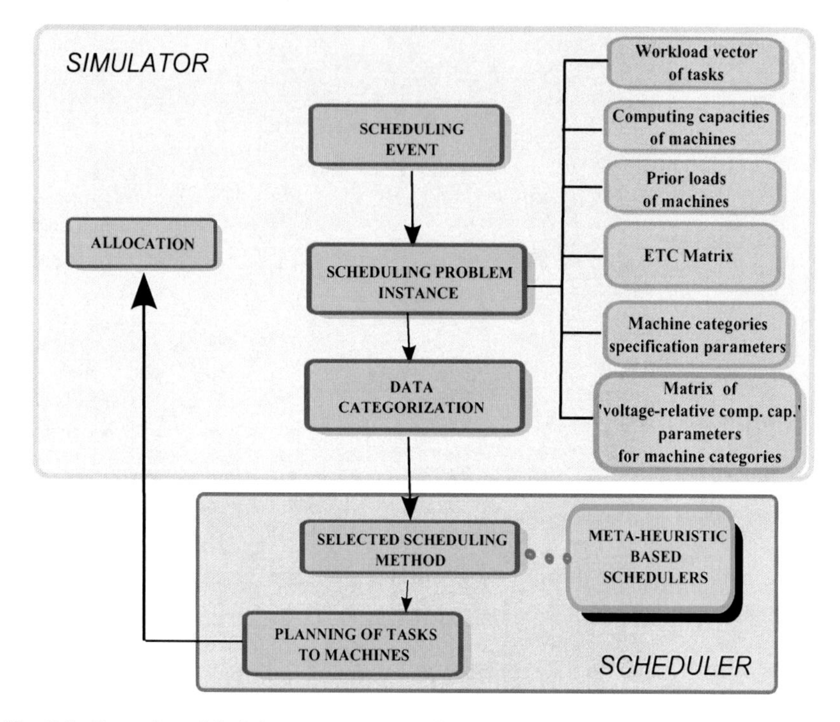

Fig. 8.1 General model of the 'energy-aware' implementation of *Sim-G-Batch* simulator

The machines can work at 16 DVFS levels and can be categorized into three 'energetic' resource classes, Class I, Class II, and Class III. The class identifiers have been selected randomly for the machines. The values of supply voltages and relative machine frequencies at all DVFS levels are specified in Table 8.1. The general settings of the simulator for four grid scenarios—*Small*, *Medium*, *Large* and *Very Large* grids—are the same as in the empirical analysis presented in Chapter 4, and defined in Table 4.1.

The configurations of key parameters for both implementations of single-population GA, *IGA* and *Green-HGS-Sched* meta-heuristics are presented in Tables 8.3, 8.4 and 8.5. The size of the initial and intermediate populations in *IGA* depends on the implementation of the genetic engine in islands and are the same as for the single-population *GA-Elit* and *GA-St* algorithms. The parameters for single-population GA schedulers and the energy-aware implementation of *HGS-Sched* are similar to the settings defined in Chapters 4 and 5 for GAs and *HGS-Sched*.

8.4.1 Energy Aware Genetic-Based Batch Schedulers

Six genetic-based meta-heuristics have been developed for minimizing the makespan and energy consumption in the **Max-Min** and **Modular Power Supply** scheduling

Table 8.2 Six GA-based grid schedulers evaluated in the empirical analysis

Scheduler	Type of algorithm	Replacement method
GA-Elit	Single-population GA	Elitist Generational
GA-St	Single-population GA	Struggle
IGA-Elit	Island GA	Elitist Generational
IGA-St	Island GA	Struggle
HGS-Elit	*Green-HGS-Sched*	Elitist Generational
HGS-St	*Green-HGS-Sched*	Struggle

modes defined in the previous section. The configuration of the genetic operators in those meta-heuristics is presented in Table 8.2.

The aforementioned methodologies differ in the implementation of the replacement mechanism in the main genetic framework. The *Elitist Generational* replacement is used in *xxx-Elit* algorithms and the *Struggle* procedure in *xxx-St* algorithms. Both single-population GAs—*GA-Elit* and *GA-St*—are implemented as the main genetic mechanism in *IGA-Elit*, *HGS-Elit*, *IGA-St*, and *HGS-St* respectively.

The concept of IGA algorithm with the specification of all key parameters for this strategy, was presented in Chapter 4 (Sec. 4.4.2.1).

The template of the main genetic engine in all schedulers is defined in Alg. 1 in Chapter 3, and the encoding methods for the schedulers are the same as in Sec. 3.3. The combination of the main operators for all schedulers is similar to the optimal configuration of the genetic mechanism in the HGS Sched generated in Chapter 4. The *Linear Ranking* is used as the selection scheme, and the *Cycle Crossover (CX)* and *Move* mutation are selected as the main genetic operators.

The configurations of key parameters for both implementations of single-population GA, *IGA* and *Green-HGS-Sched* meta-heuristics are presented in Tables 8.3, 8.4 and 8.5. The size of the initial and intermediate populations in *IGA* depends on the implementation of the genetic engine in islands and are the same as for the single-population *GA-Elit* and *GA-St* algorithms. The parameters for single-population GA schedulers and the energy-aware implementation of *HGS-Sched* are similar to the settings defined in Chapters 4 and 5 for GAs and *HGS-Sched*.

The relative performance of all six schedulers has been quantified with the following two metrics:

- minimal makespan defined as follows:

$$makespan = min\{Makespan_I, Makespan_{II}\} \tag{8.21}$$

Table 8.3 GA setting for static and dynamic benchmarks

Parameter	GA-Elit	GA-St
evolution steps	$5*m$	$20*m$
pop. size (pop_size)	$\lceil(\log_2(m))^2 - \log_2(m)\rceil$	$4*(\log_2(m)-1)$
intermediate pop.	$pop\_size - 2$	$(pop\_size)/3$
cross probab.	1.0	1.0
mutation probab.	0.2	
max_time_to_spend	30 secs $(static)$ / 45 secs $(dynamic)$	

Table 8.4 *HGS-Sched* settings for static and dynamic benchmarks

Parameter	
period_of_metaepoch	$20*n$
nb_of_metaepochs	10
degrees of branches (t)	0 and 1
population size in the core	$3*(\lceil 4*(\log_2 n - 1)/(11.8)\rceil)$
population size in the sprouted branches (b_pop_size)	$(\lceil(4*(\log_2 n - 1)/(11.8)\rceil)$
intermediate pop. in the core	$abs((r\_pop\_size)/3)$
intermediate pop. in the sprouted branch	$abs((b\_pop\_size)/3)$
cross probab.	0.9
mutation probab. in core	0.4
mutation probab. in the sprouted branches	0.2
max_time_to_spend	40 secs $(static)$ / 70 secs $(dynamic)$

- a relative energy consumption improvement rate expressed as follows:

$$Im(E) = \frac{E_I - E_{batch}}{E_{batch}} \cdot 100\%, \tag{8.22}$$

where E_{II} and E_I are defined in Eq. (8.10) and Eq. (8.17) respectively;

Table 8.5 Configuration of *IGA* algorithm

Parameter	
it_d	$20 * n$
mig	5 %
number of islands (demes)	10
cross probab.	1.0
mutation probab.	0.2
$max\_time\_to\_spend$	40 secs ($static$) / 70 secs ($dynamic$)

8.4.2 Results

Each experiment has been repeated 30 times under the same configuration of operators and parameters. The box-plots of the first, mean, and the third quantiles (confidence level - 95 %) for the makespan and relative energy consumption rate $Im(E)$ are presented in Fig. 8.2–8.5.

Makespan optimization results

Fig. 8.2–8.5 depicts the box-plots of the makespan values for six considered schedulers. The makespan is measured and expressed in arbitrary time units defined for the execution of tasks.

Both implementations of the *Green-HGS-Sched* have achieved the best results in all instances but *Large grid* in the static case and *Small* and *Large* instances in the dynamic case, where they are outperformed by the *IGA* algorithm.

On the one hand, a simple comparison of the impact of the replacement method on the algorithms performance provided for all pairs of the *xxx-Elit* and *xxx-St* schedulers shows that *Struggle* replacement is much more effective that *Elitist Generational* method in the case of single-population *GA* and *IGA* schedulers. It also confirms the results of the preliminary study on the effectiveness of single-population genetic schedulers in CGs presented in [82].

On the other hand, for the *Green-HGS-Sched* the situation is completely different. In most of the scheduling instances the effectiveness of both hierarchical schedulers are at comparative levels, with little advantage for the elite technique in the dynamic cases. It seems to indicate that the replacement mechanism does not play a crucial role in the fast exploration of the search space by the *Green-HGS-Sched*. Such exploration process can be construed as very slow when using the conventional GA and IGA schedulers. The core of *Green-HGS-Sched* can activate the more accurate processes in the neighborhoods of the partial solutions of the problem. Those solutions

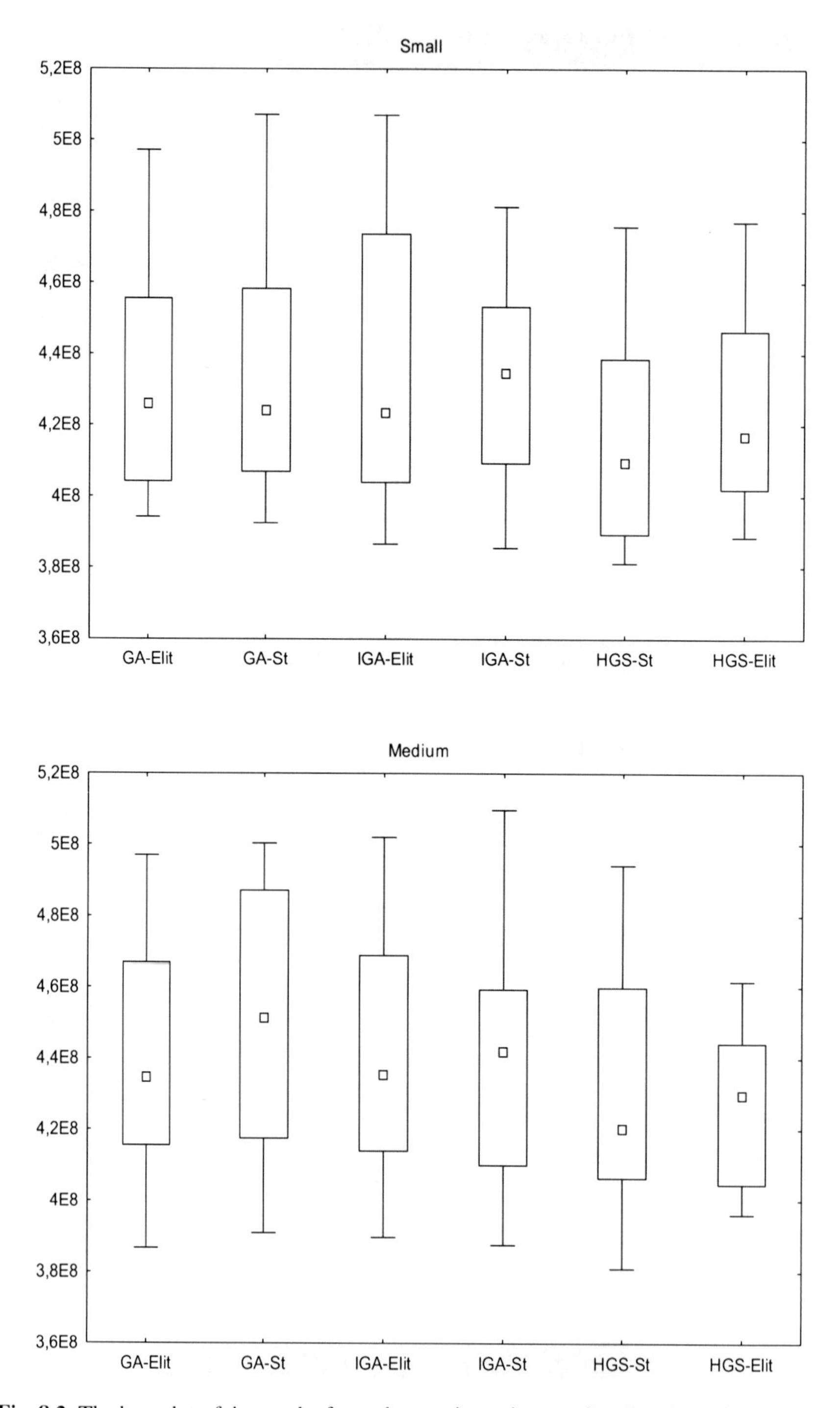

Fig. 8.2 The box-plot of the results for makespan in static case: Small and Medium grids

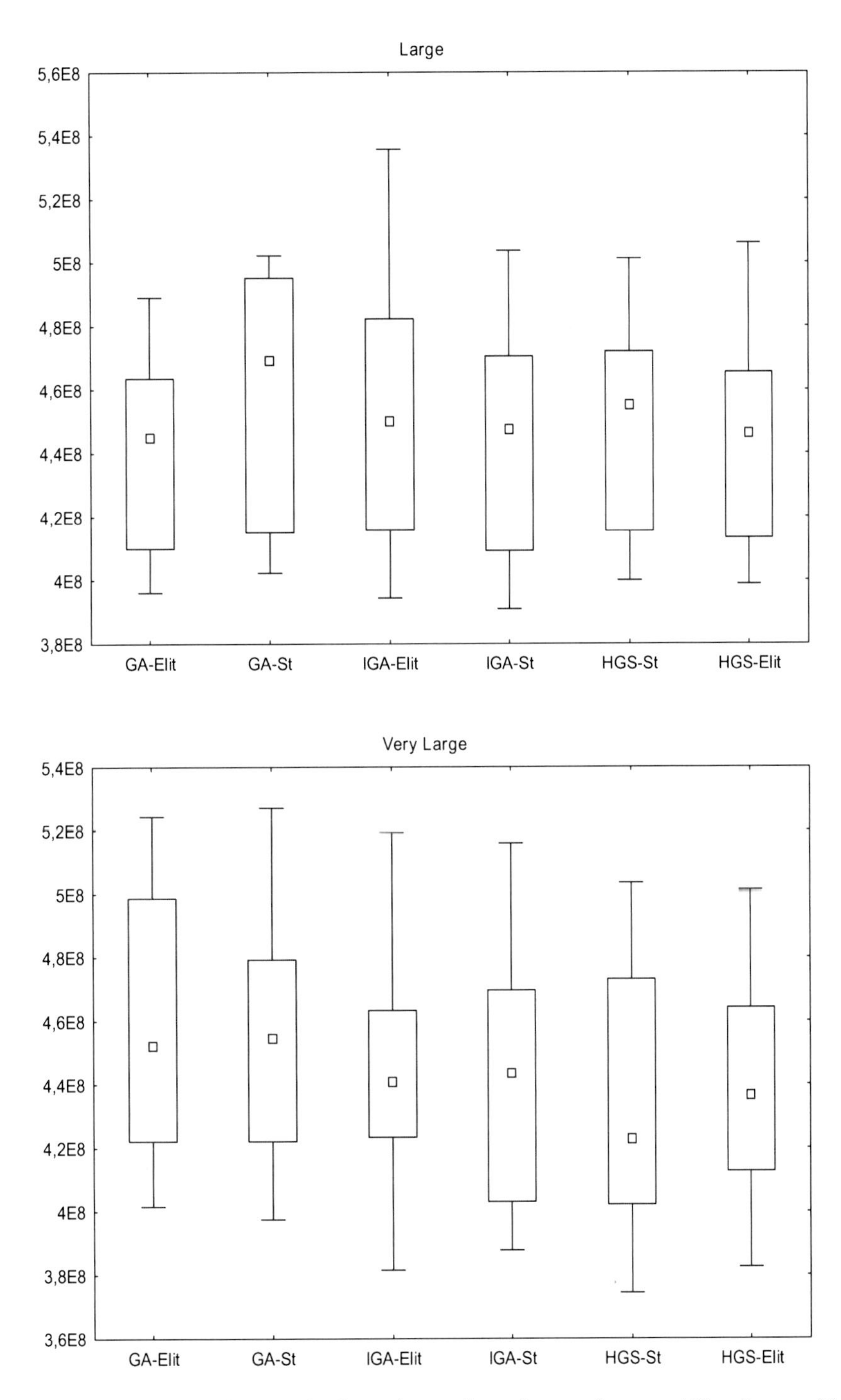

Fig. 8.3 The box-plot of the results for makespan in static case: Large and Very Large grids

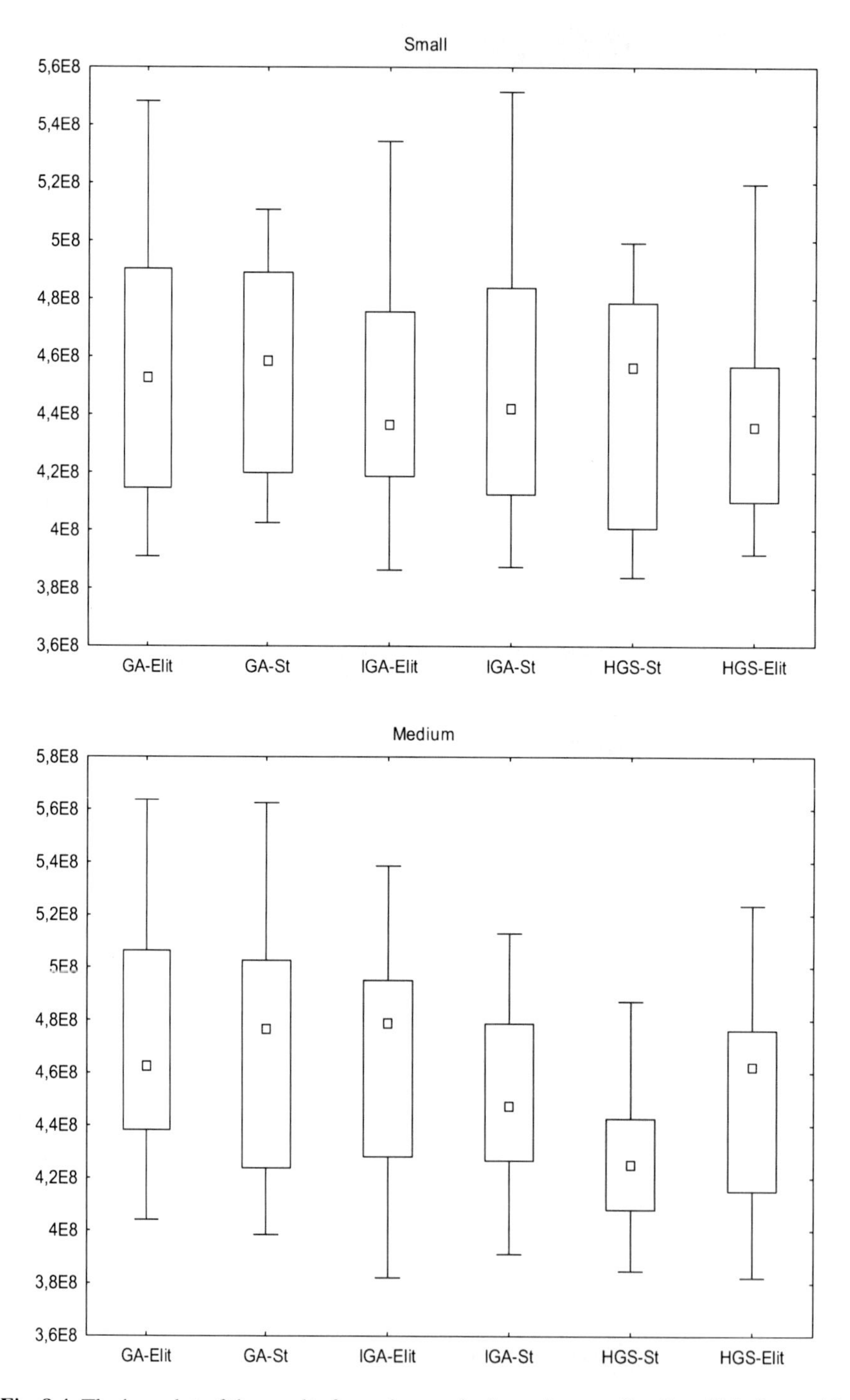

Fig. 8.4 The box-plot of the results for makespan in dynamic case: Small and Medium grids

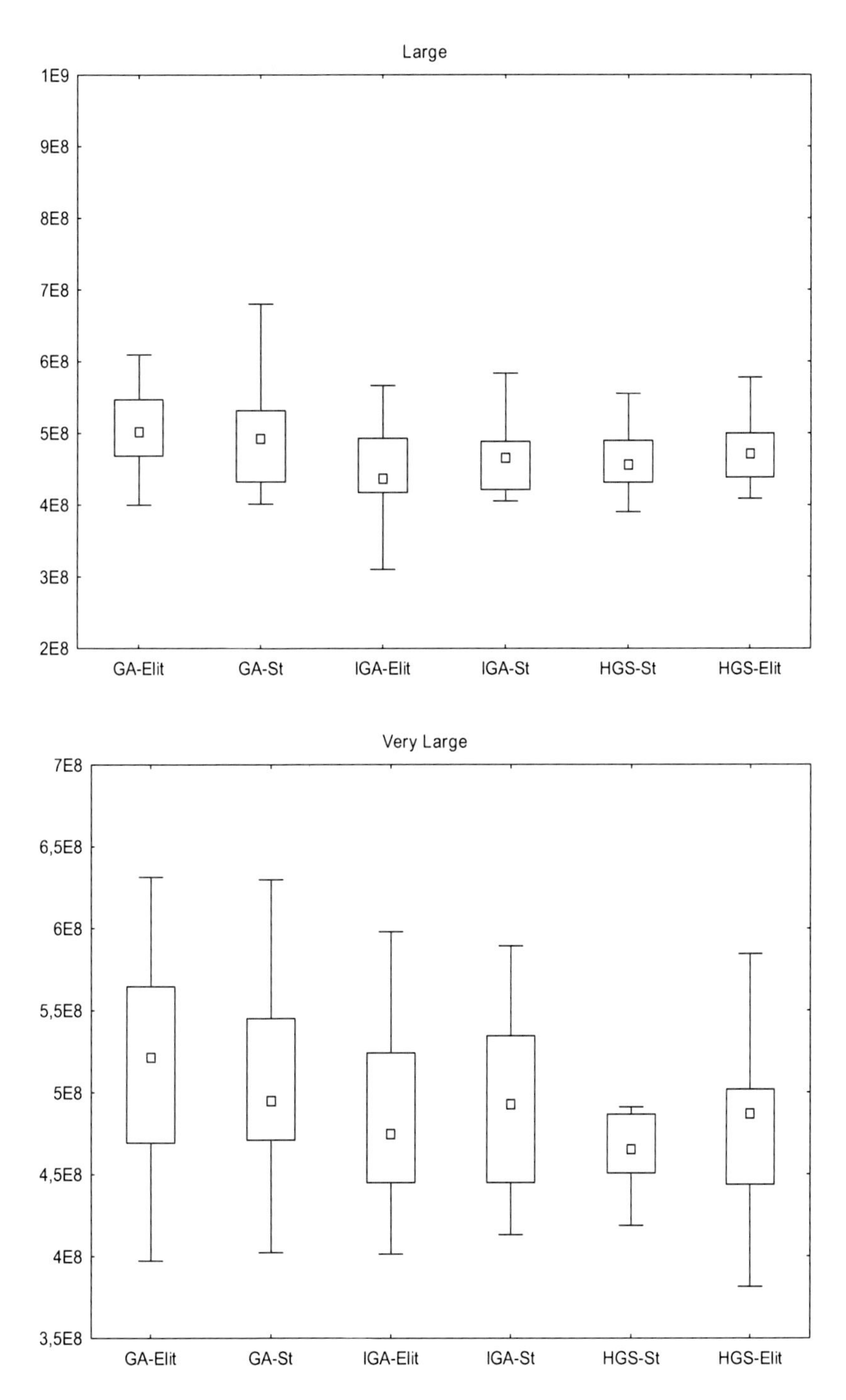

Fig. 8.5 The box-plot of the results for makespan in dynamic case: Large and Very Large grids

may be not detected by the other schedulers, which makes the *Green-HGS-Sched* very effective in the exploration of new regions in the optimization domain and in escaping the basins of attraction of the local solutions.

The complexity of the hierarchic system is, in fact, not a drawback of the *Green-HGS-Sched*, because the constraints of the execution time for HGS and IGA are exactly the same. The ranges in the achieved makespan values for all considered meta-heuristics are not greater than 30 – 35 % of the mean makespan values, which means that the stability of all schedulers in all cases is at an acceptable level. The distributions of the makespan results are asymmetric: the skewness in the static case is positive—for GA and IGA and negative for *Green-HGS-Sched* in most of the static instances—and it is negative in the dynamic grids for almost all schedulers. It also implies that the reduction of the average makespan in this case is more difficult than in the static case, which confirms the complexity of the problem in the realistic dynamic grid scenarios.

Energy optimization results

The main effect of the makespan minimization is arguably the balance of the loads in grid resources. The application of the DVFS technique typically leads to a significant reduction of the energy consumption in the system, especially in the case of substantial differences in the loads of particular machines. The box-plots for the energy saving rates $Im(E)$ are presented in Fig. 8.6–8.9.

The results of the energy optimization are slightly different in comparison with the makespan ones. In this case, each of the *IGA-Elit* and *GA-Elit* algorithms outperforms the rest of the schedulers in five instances, and the single-population GAs are the best in the three cases. *Green-HGS-Sched* is not as effective in energy optimization as in the makespan minimization. It means that this algorithm works quite well in **Min-Max** scenario: the makespan is relatively short, and the scaling of the voltage supply may lead to not so significant energy conservation.

In the case of single population and island models, the extensions of the completion times of the tasks in **Modular Power Supply** mode allow to keep the machines busy for a longer time than in the **Min-Max** mode. However, the average difference in energy saving rates achieved by the *Green-HGS-Sched* scheduler and the remaining meta-heuristics does not exceed 10 %, which is smaller than in the makespan case (15 %). This signifies that the average cumulative energy utilization achieved by *Green-HGS-Sched* is lower than the island GA and conventional GAs. The range of the average energy saving rate values is 10%–35% for most of the schedulers. It can also be observed that the skewness of the distribution of the results is positive or neutral for the worst 'energy optimizers' and negative for the best ones.

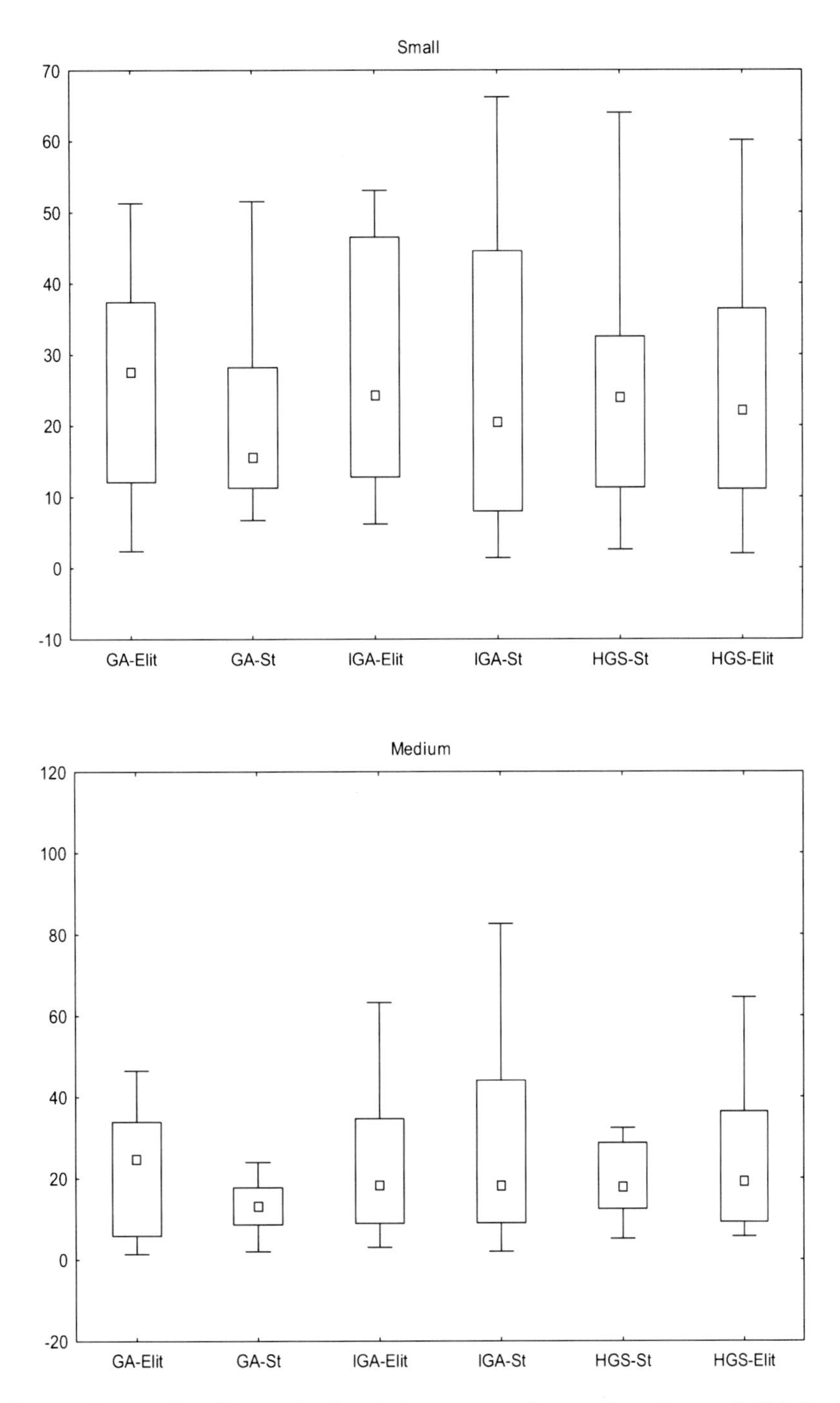

Fig. 8.6 The box-plot of the results for relative energy saving rate in static case (in %): Small and Medium grids

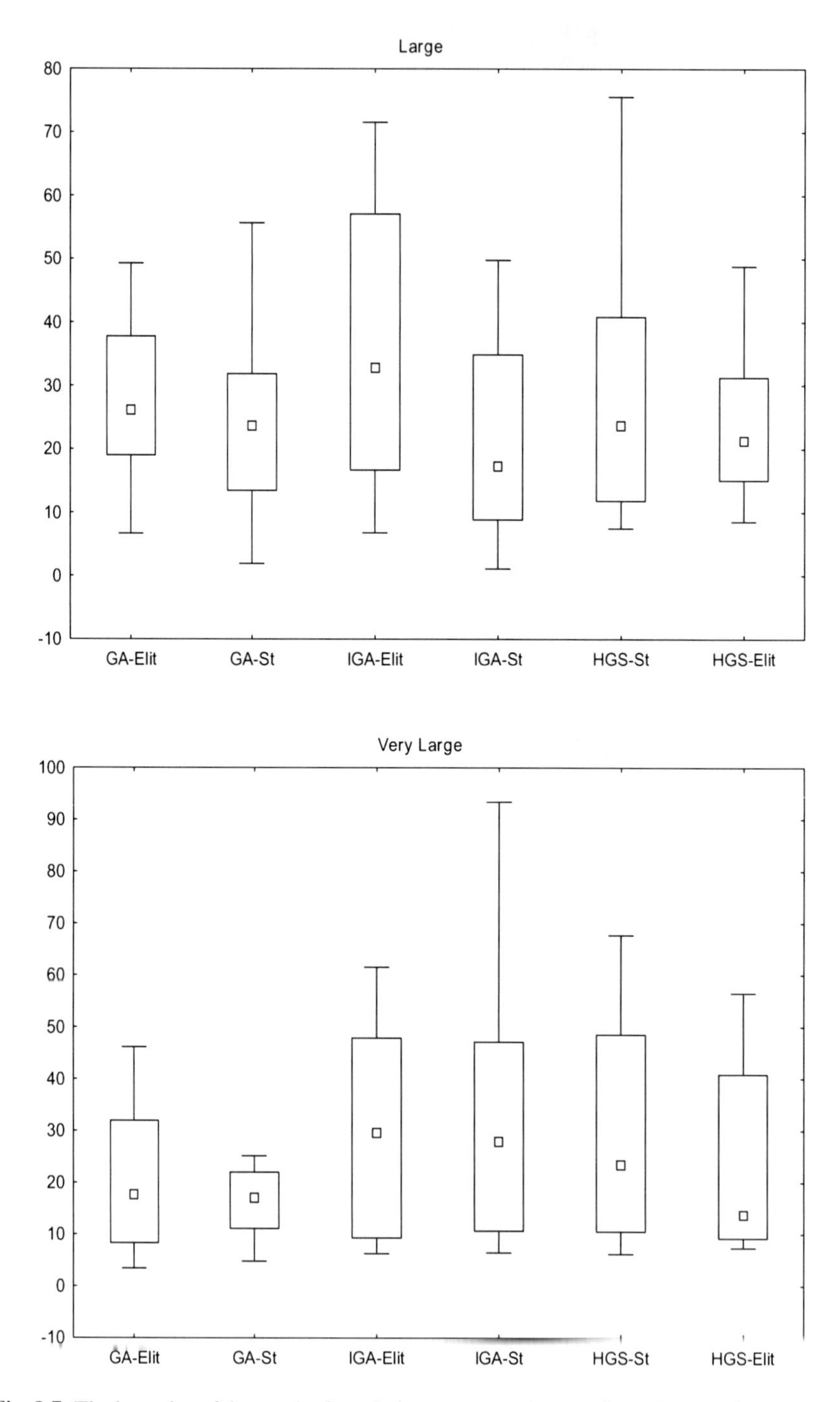

Fig. 8.7 The box-plot of the results for relative energy saving rate in static case (in %): Large and Very Large grids

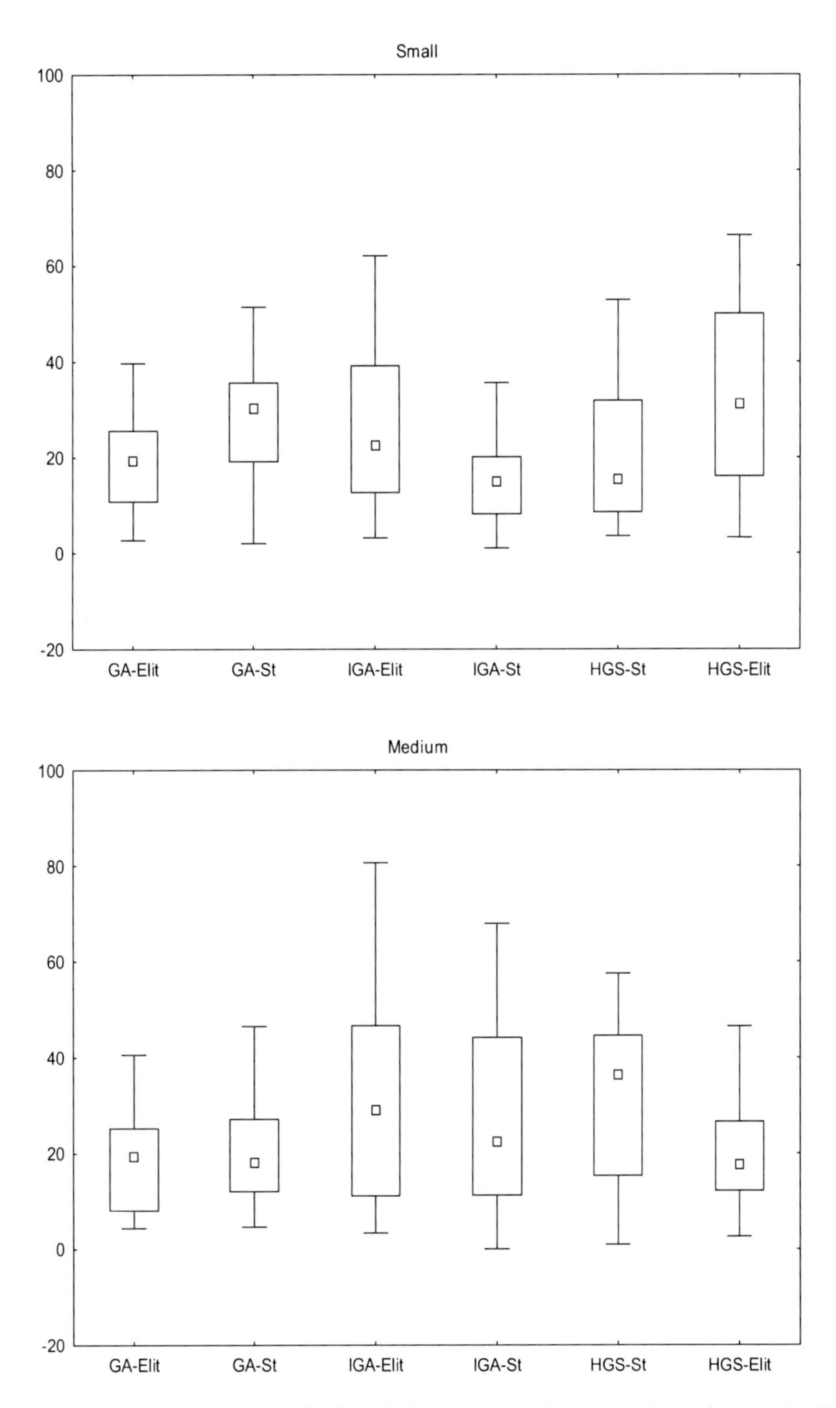

Fig. 8.8 The box-plot of the results for relative energy saving rate in dynamic case (in %): Small and Medium grids

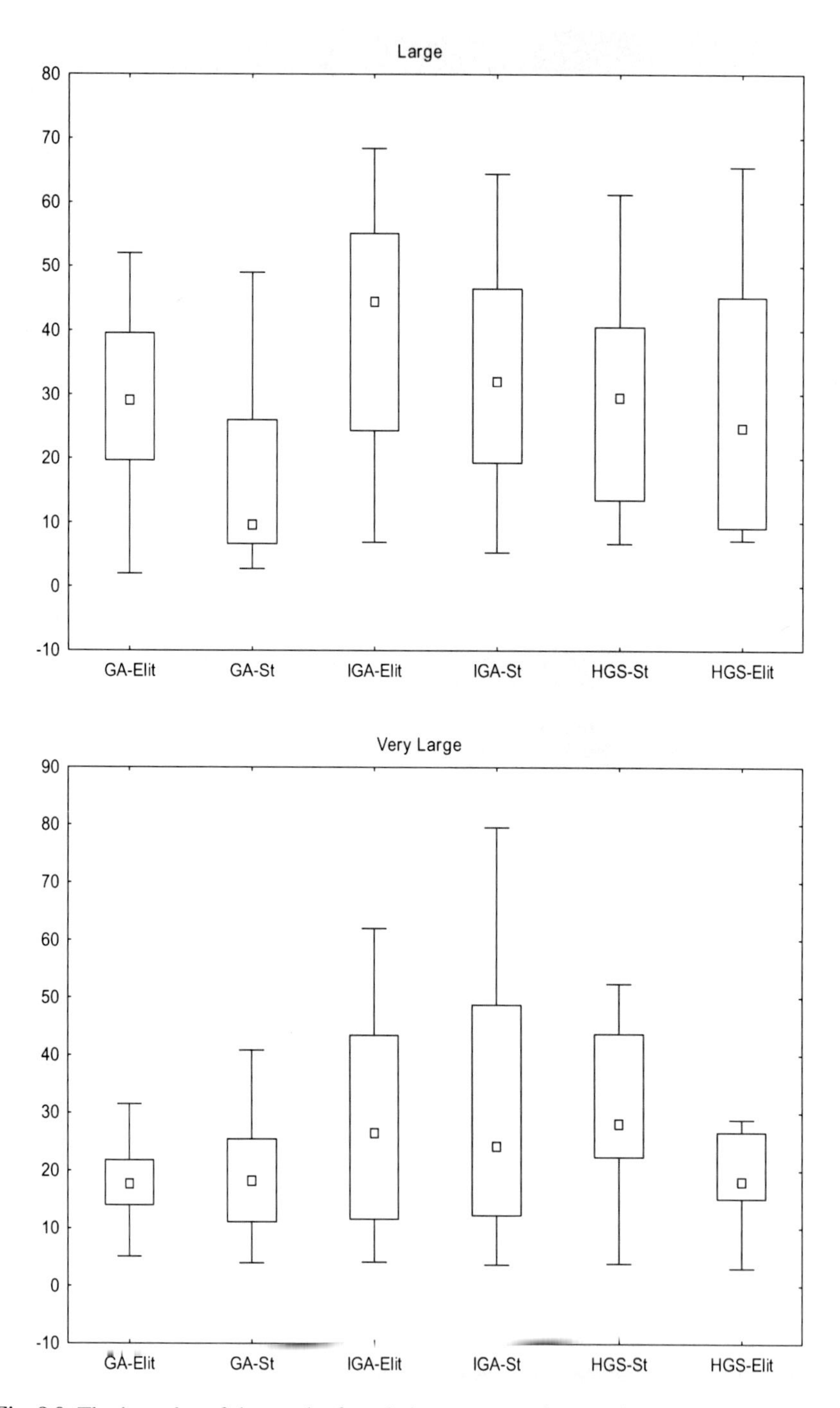

Fig. 8.9 The box-plot of the results for relative energy saving rate in dynamic case (in %): Large and Very Large grids

8.5 Summary

This chapter addressed the problem of optimizing the energy utilized in CGs in independent batch scheduling. The energy management model is based on *Dynamic Voltage Scaling (DVFS)* technique adapted to the dynamic Grid environment. The energy-aware Grid scheduling was formalized as a bi-objective optimization problem with makespan and average cumulative energy consumption as the main objectives.

For solving the addressed Grid scheduling problem, two implementations of an energy-efficient Hierarchical Grid Scheduler *HGS-Sched* were developed evaluated in two 'energetic' scheduling modes in static and dynamic Grid scenarios under the makespan and relative energy consumption improvement rate criteria. The efficiencies of the hierarchical schedulers were compared with the results achieved by four single-population Genetic Algorithm (GA) and Island GA schedulers. The simulation results confirmed the effectiveness of the proposed schedulers in the reduction of the energy consumed by the whole system and in dynamic load balancing of the resources in Grid clusters, which is sufficient to maintain the desired quality level(-s).

Summary

The concepts of today's grid computing systems grown far beyond the original model of Ian Foster of the power electric grid. The modern large-scale grids are made up of hundreds or thousands of various components (computers, databases, etc), not just the computing nodes and high performance computing platforms. Due to the high heterogeneity of the users and resources, the grid managers in one locality (geographical or managerial) might not be able to have control over other parts of the system. The ability of the efficient scheduling of the grid applications and the allocation of the resources in a desired configuration of all system components in a scalable and robust manner is essential in today's grid computing.

The categorization of the grid scheduling problems presented in this book allows to look at the old scheduling models from a contemporaneous and unique perspective. Two new scheduling criteria, namely security and energy consumption, usually considered as the separate optimization problems, are embedded in the proposed scheduling models. The simulated grid scenarios in such cases can better illustrate the realistic systems, in which large number of variables, numerous objectives, constraints, and business rules, all contributing in various ways must be analyzed.

In many cases the management system in the grid environment should be able to group, predict, and classify the different sets of individual rules and requirements of the grid users. Therefore the scheduling problem in grids has been interpreted in this book as a difficult decision problem for grid users working at different levels of the system. Users decisions and fundamental features arising in the users' behavior, such as cooperativeness, trustfulness and symmetric and asymmetric roles, are modeled based on the game theory paradigm.

The main reason behind the complexity of the multi-criteria grid scheduling is that this problems consist of several interconnected components (criteria, sub-problems), which makes many standard approaches ineffective. Even if the exact and efficient algorithms for solving particular components or aspects of an overall problem are well-known, these algorithms only yield solutions to sub-problems, and it remains an open question how to integrate these partial solutions to achieve a global optimum. Metaheuristics, due to their robustness and high scalability, are able to tackle the various and also sometimes conflicting scheduling attributes and criteria. A generic model for a hierarchical multi-population genetic scheduler presented in this book, enables an undemanding configuration of the numerous genetic operators and an effective exploration of the search space with an adaptive accuracy. This model has been easily adapted to a range of scheduling scenarios. The functionality of this model and its effectiveness in multi-criteria grid scheduling have been justified in the comprehensive experimental analysis.

All models presented in this book are in fact not restricted just to the conventional grid systems. They may be easily adapted to cloud environments, where security awareness and intelligent power management are the hottest research issues.

References

1. Abawajy, J.: An efficient adaptive scheduling policy for high performance computing. Future Generation Computer Systems 25(3), 364–370 (2009)
2. Abraham, A., Buyya, R., Nath, B.: Nature's heuristics for scheduling jobs on computational grids. In: Proc. of the 8th IEEE International Conference on Advanced Computing and Communications, India, pp. 45–52 (2000)
3. Abraham, A., Jainb, R., Thomasc, J., Hana, S.: D-scids: Distributed soft computing intrusion detection system. Journal of Network and Computer Applications 30(1), 81–98 (2007)
4. Alba, E., Troya, J.: A survey of parallel distributed genetic algorithms. Complexity 4(4), 31–52 (1999)
5. Ali, S., Siegel, H., Maheswaran, M., Ali, S., Hensgen, D.: Task execution time modeling for heterogeneous computing systems. In: Proc. of the Workshop on Heterogeneous Computing, pp. 185–199 (2000)
6. Ali, S., Siegel, H., Maheswaran, M., Hensgen, D., Ali, S.: Representing task and machine heterogeneities for heterogeneous computing systems. Tamkang Journal of Science and Engineering 3(3), 195–207 (2000)
7. Armstrong, R., Hensgen, D., Kidd, T.: The relative performance of various mapping algorithms is independent of sizable varie,nces in run-time predictions. In: Proc. of the 7th IEEE Heterogeneous Computing Workshop (HCW 1998), pp. 79–87 (1998)
8. Azzedin, F., Maheswaran, M.: Integrating trust into grid resource management systems. In: Proc. Int. Conf. Parallel Processing (2002)
9. Azzemi, N.: A multiobjective evolutionary approach for constrained joint source code optimization. In: Proc. of the ISCA 19th International Conference on Computer Application in Industry (CAINE 2006), Las Vegas, Nevada, USA, pp. 175–180 (2006)
10. Baçsar, T., Olsder, G.: Dynamic Non-cooperative Game Theory, 2nd edn. Academic Press, London (1995)
11. Back, T., Fogel, D., Michalewicz, Z.E.: Volutionary Computation, Part 1 and 2. IOP Publishing Ltd. (2000)
12. Baker, M., Buyya, R., Laforenza, D.: Grids and grid technologies for wide-area distributed computing. Software–Practice and Experience 32(15), 1437–1466 (2002)
13. Baker, R.J.: CMOS: circuit design, layout, and simulation, 2nd edn. Wiley (2008)
14. Bartschi Wall, M.: A Genetic Algorithm for Resource-Constrained Scheduling. PhD Thesis. Massachusetts Institute of Technology, MA (1996)
15. Beasley, J.: Or-library: distributing test problems by electronic mail. Journal of the Operational Research Society 41(11), 1069–1072 (1990)

16. Błażewicz, J., Brauner, N., Finke, G.: Scheduling with discrete resource constraints. In: Leung, J.Y.-T. (ed.) Handbook of Scheduling, ch. 23, pp. 23.1–23.18. Chapman & Hall/CRC (2004)
17. Błażewicz, J., Lenstra, J., Rinnooy Kan, A.: Scheduling subject to resource constraints: classification and complexity. Discrete Applied Mathematics 5(1), 11–24 (1983)
18. Bogdański, M., Kołodziej, J., Xhafa, F.: Supporting the security awareness of ga-based grid schedulers by artificial neural networks. In: Proc. of International Conference on Complex, Intelligent, and Software Intensive Systems (CISIS 2011), Seoul, June 30-July 2, pp. 277–284. IEEE Soc. Press, Los Alamitos (2011)
19. Brandic, I., Pllana, S., Benkner, S.: Specification, planning, and execution of qos-aware grid workflows. In: Market Oriented Grid and Utility Computing. Wiley (2009)
20. Braunt, T., Siegel, H., Beck, N., Boloni, L., Maheswaran, M., Reuther, A., Robertson, J., Theys, M., Yao, B., Hensgen, D., Freund, R.: A comparison of eleven static heuristics for mapping a class of independent tasks onto heterogeneous distributed computing systems. J. of Par. and Dist. Computing 61(6), 810–837 (2001)
21. Brucker, P.: Scheduling Algorithms. Springer (2007)
22. Brucker, P., Drexl, A., Mohring, R., Neumann, K., Pesch, E.: (1999)
23. Buyya, R.: The world-wide grid: `http://www.buyya.com/ecogrid/wwg/`
24. Buyya, R.: Economic-based Distributed Resource Management and Scheduling for Grid Computing. PhD Thesis. Monash University, Australia (2002)
25. Buyya, R., Abramson, D., Giddy, J.: Nimrod/g: An architecture for a resource management and scheduling system in a global computational grid. In: Proc. of the Int. Conference on High Performance Computing in Asia-Pacific Region, HPC Asia 2000 (2000)
26. Buyya, R., Bubendorfer, K.E.: Market Oriented Grid and Utility Computing. Wiley (2009)
27. Buyya, R., Murshed, M.: Gridsim: A toolkit for the modeling and simulation of distributed resource management and scheduling for grid computing. Concurrency and Computation: Practice and Experience 14(13-15), 1175–1220 (2002)
28. Cao, X., Zhang, H., Shi, J., Cui, G.: Cluster heads election analysis for multi-hop wireless sensor networks based on weighted graph and particle swarm optimization. In: Proc. of the 4th International Conference on Natural Computation (ICNC), vol. 7, pp. 599–603 (2008)
29. Carretero, J., Xhafa, F.: Using genetic algorithms for scheduling jobs in large scale grid applications. Journal of Technological and Economic Development –A Research Journal of Vilnius Gediminas Technical University 12(1), 11–17 (2006)
30. Chang, R., Chang, J., Lin, P.: An ant algorithm for balanced job scheduling in grids. Future Generation Computer Systems 25(2009), 20–27 (2009)
31. Chen, X., Ong, Y., Lim, M.H., Tan, K.: A multi-facet survey on memetic computation. IEEE Transactions on Evolutionary Computation 15(5), 591–607 (2011)
32. Chervenak, A., Foster, I., Kesselman, C., Salisbury, C., Tuecke, S.: Thedatagrid: Towards anarchitecture for the distributed management and analysis of large scientific datasets. Journal of Network and Computer Applications 23(3), 187–200 (2000)
33. Cirne, W., Brasileiro, F., Andrade, N., Costa, L., Andrade, A., Novaes, R., Mowbray, M.: Labs of the world, unite!!! Journal of Grid Computing 4(3), 225–246 (2006)
34. Cody, E., Sharman, R., Rao, R.H., Upadhyaya, S.: Security in grid computing: A review and synthesis. Decision Support Systems 44, 749–764 (2008)
35. Cohen, J., Cordeiro, D., Trystram, D., Wagner, F.: Multi-organization scheduling approximation algorithms. Concurrency and Computation: Practice and Experience 23(17), 2220–2234 (2011)

36. Davis, L.E.: Handbook of Genetic Algoriithms. Van Nostrand Reinhold, New York (1991)
37. Di Martino, V., Mililotti, M.: Sub-optimal scheduling in a grid using genetic algorithms. Parallel Computing 30, 553–565 (2004)
38. Dorigo, M., Stützle, T.: Ant Colony Optimization. MIT Press (2004)
39. Edlefsen, L., Millham, C.: On a formulation of discrete n-person non-cooperative games. Metrika 18(1), 31–34 (1972)
40. EGI: http://knowledge.eu-egi.eu/knowledge/index.php
41. Ehrgott, M.: Multicriteria Optimization. Springer (2005)
42. El-Mihoub, T., Hopgood, A., Nolle, L., Battersby, A.: Hybrid genetic algorithms: A review. Engineering Letters 13, 124–137 (2006)
43. Enokido, T., Aikebaier, A., Takizawa, M.: Process allocation algorithms for saving power consumption in peer-to-peer systems. IEEE Tran. on Industrial Electronics 58(6), 2097–2105 (2011)
44. Feller, E., Rilling, L., Morin, C.: Energy-aware ant colony based workload placement in clouds. In: INRIA Report RR-7622, Rennes, France (2011)
45. Fibich, P., Matyska, L., Rudová, H.: Model of grid scheduling problem (2005)
46. Fonseca, C.M., Fleming, P.J.: An overview of evolutionary algorithms in multiobjective optimization. Evolutionary Computation 3(1), 1–16 (1995)
47. Foster, I., Kesselman, C.: Globus: A metacomputing infrastructure toolkit. Int. J. Supercomputer Applications 11(2), 115–128 (1997)
48. Foster, I., Kesselman, C.: The Grid: Blueprint for a New Computing Infrastructure. Morgan-Kaufmann (1998)
49. Gao, Y., Rong, H., Huang, J.: Adaptive grid job scheduling with genetic algorithms. Future Generation Computer Systems 21(1), 151–161 (2005)
50. Garg, S., Buyya, R., Segel, H.: Scheduling parallel aplications on utility grids: Time and cost trade-off management (2009)
51. Garg, S., Yeo, C., Anandasivam, A., Buyya, R.: Energy-Efficient Scheduling of HPC Applications in Cloud Computing Environments. CoRR abs/0909.1146 (2009)
52. Ghosh, P., Roy, N., Basu, K., Das, S.: A game theory based pricing strategy for job allocation in mobile grids. In: Proc. of the 18th IEEE International Parallel and Distributed Processing Symposium (IPDPS 2004), Santa Fe, New Mexico (2004)
53. Glover, F., Laguna, M.: Tabu Search. Kluwer Academic Publishers (1997)
54. Goldberg, D.: Genetic algorithms in search, optimization and machine learning. Addison-Wesley (1989)
55. Goldberg, D., Lingle, R.: Alleles, loci and the travelling salesman problem. In: Proc. of the ICA 1985, Pittsburgh (1985)
56. Graham, R., Lawler, E., Lenstra, J., Rinnooy Kan, A.: Optimization and approximation in deterministic sequencing and scheduling: a survey. Annals of Discrete Mathematics 5, 287–326 (1979)
57. Grid, I.: http://www-1.ibm.com/grid
58. Grid, S.D.: http://www.desktopgrid.hu
59. Grueninger, T.: Multimodal optimization using genetic algorithms. Technical report. Department of Mechanical Engineering, MIT, Cambridge, MA (1997)
60. Guzek, K., Pecero, J.E., Dorrosoro, B., Bouvry, P., Khan, S.U.: A cellular genetic algorithm for scheduling applications and energy-aware communication optimization. In: Proc. of the ACM/IEEE/IFIP Int. Conf. on High Performance Computing and Simulation (HPCS), Caen, France, pp. 241–248 (2010)
61. Haykin, S.: Neural Networks: A Comprehensive Foundation, 2nd edn. Prentice-Hall (1999)

62. Hernández, H., Blum, C., Francès, G.: Ant Colony Optimization for Energy-Efficient Broadcasting in Ad-Hoc Networks. In: Dorigo, M., Birattari, M., Blum, C., Clerc, M., Stützle, T., Winfield, A.F.T. (eds.) ANTS 2008. LNCS, vol. 5217, pp. 25–36. Springer, Heidelberg (2008)
63. Holland, J.: Adaptation in Natural and Artificial Systems. University of Michigan Press (1975)
64. Hotovy, S.: Workload evolution on the cornell theory center IBM SP2. In: Proc. of the Workshop on Job Scheduling Strategies for Parallel, IPPS 1996, pp. 27–40 (1996)
65. Humphrey, M., Thompson, M.: Security implications of typical grid computing usage scenarios. In: Proc. of the Conf. on High Performance Distributed Computing (2001)
66. Hwang, K., Xu, Z.: Parallel Computing: Technology, Architecture, Programming. McGraw-Hill (2009)
67. Hwang, S., Kesselman, C.: A flexible framework for fault tolerance in the grid. J. of Grid Computing 1(3), 251–272 (2003)
68. Johnston, W., Gannon, D., Nitzberg, B.: Grids as production computing environments: The engineering aspects of nasa's information power grid. In: Proc. of the 8th IEEE International Symposium on High Performance Distributed Computing, Redondo Beach, CA. IEEE Computer Society Press, Los Alamitos (1999)
69. Kacprzyk, J.: Multistage Decision Making under Fuzziness. Verlag TUV, Rheinland (1983)
70. Katoen, J., Khattri, M., Zapreev, I.S.: A markov reward model checker. In: Proc. of the QEST:International Conference on the Quantitative Evaluation of Systems, pp. 243–244. IEEE Computer Society (2005)
71. Kennedy, J., Eberhart, R.: Particle swarm optimization. In: Proc. of the IEEE International Conference on Neural Networks, November 27-December 1, vol. 4 (1995)
72. Kessaci, Y., et al.: Parallel Evolutionary Algorithms for Energy Aware Scheduling. In: Bouvry, P., González-Vélez, H., Kołodziej, J. (eds.) Intelligent Decision Systems in Large-Scale Distributed Environments. SCI, vol. 362, pp. 75–100. Springer, Heidelberg (2011)
73. Khan, S.: A goal programming approach for the joint optimization of energy consumption and response time in computational grids. In: Proc. of the 28th IEEE International Performance Computing and Communications Conference (IPCCC), Phoenix, AZ, USA, pp. 410–417 (2009)
74. Khan, S.: A self-adaptive weighted sum technique for the joint optimization of performance and power consumption in data centers. In: Proc. of the 22nd International Conference on Parallel and Distributed Computing and Communication Systems (PDCCS), USA, pp. 13–18 (2009)
75. Khan, S., Ahmad, I.: Non-cooperative, semi-cooperative, and cooperative games-based grid resource allocation. In: Proceedings of International Parallel and Distributed Proceedings Symposium (IPDPS 2006), pp. 101–104 (2006)
76. Khan, S.U., Ahmad, I.: A cooperative game theoretical technique for joint optimization of energy consumption and response time in computational grids. IEEE Tran. on Parallel and Distributed Systems 20(3), 346–360 (2009)
77. Kliazovich, D., Bouvry, P., Khan, S.U.: Dens: Data center energy-efficient network-aware scheduling. In: Proc. of ACM/IEEE International Conference on Green Computing and Communications (GreenCom), Hangzhou, China, pp. 69–75 (December 2010)
78. Klusaček, D., Rudová, H.: Efficient grid scheduling through the incremental schedule-based approach. Computational Intelligence 27(1), 4–22 (2011)
79. Kolodziej, J.: Modelling Hierarchical Genetic Strategy as a Family of Markov Chains. In: Wyrzykowski, R., Dongarra, J., Paprzycki, M., Waśniewski, J. (eds.) PPAM 2001. LNCS, vol. 2328, pp. 595–599. Springer, Heidelberg (2002)

80. Kołodziej, J., Gwizdała, R., Wojtusiak, J.: Hierarchical genetic strategy as a method of improving search efficiency. In: Schaefer, R., Sędziwy, S. (eds.) Advances in Multi-Agent Systems, pp. 149–161. UJ Press, Cracow (2001)
81. Kołodziej, J., Jakubiec, W., Starczak, M., Schaefer, R.: Hierarchical genetic strategy applied to the problem of the coordinate measuring machine geometrical errors. In: Proc. of the IUTAM 2002 Symposium on Evolutionary Methods in Mechanics, September 24-27, pp. 22–30. Kluver Ac. Press, Cracow (2002)
82. Kołodziej, J., Khan, S., Xhafa, F.: Genetic algorithms for energy-aware scheduling in computational grids. In: Proc. of the 6th IEEE International Conference on P2P, Parallel, Grid, Cloud and Internet Computing (3PGCIC 2011), Barcelona, Spain, October 26-28, pp. 17–24 (2011)
83. Kołodziej, J., Rybarski, M.: An application of hierarchical genetic strategy in sequential scheduling of permutated independent jobs. In: Arabas, J. (ed.) Evolutionary Computation and Global Optimization. Lectures on Eletronics, vol. 1, pp. 95–103. Warsaw University of Technology (2009)
84. Kołodziej, J., Scheafer, R., Paszyńska, A.: Hierarchical genetic computation in optimal design. Journal of Theoretical and Applied Mechanics 42(3), 519–539 (2004)
85. Kołodziej, J., Xhafa, F.: A game-theoretic and hybrid genetic meta-heuristic model for security-assured scheduling of independent jobs in computational grids. In: Barolli, L., Xhafa, F., Venticinque, S. (eds.) Proc. of CISIS 2010, Cracow, February 15-18, pp. 93–100 (2010)
86. Kołodziej, J., Xhafa, F.: Enhancing the genetic-based scheduling in computational grids by a structured hierarchical population. Future Generation Computer Systems 27, 1035–1046 (2011)
87. Kołodziej, J., Xhafa, F.: Meeting security and user behaviour requirements in grid scheduling. Simulation Modelling Practice and Theory 19(1), 213–226 (2011)
88. Kołodziej, J., Xhafa, F.: Modelling of user requirements and behavior in computational grids. In: Proc. of 3PGCIC 2010, Fukuoka, Japan, November 4-6, pp. 548–553 (2011)
89. Kołodziej, J., Xhafa, F.: Modern approaches to modelling user requirements on resource and task allocation in hierarchical computational grids. Int. J. on Appled Mathematics and Computer Science 21(2), 243–257 (2011)
90. Kołodziej, J., Xhafa, F., Kolanko, L.: Hierarchic genetic scheduler of independent jobs in computational grid environment. In: Otamendi, J., Bargieła, A., Montes, J.L., Doncel Pedrera, L.M. (eds.) Proc. of the 23rd ECMS, Madrid, June 9-12, pp. 108–115 (2009)
91. Krauter, K., Buyya, R., Maheswaran, M.: A taxonomy of grid resource management systems for distributed computing. Software-Practice and Experience 32(2), 135–164 (2002)
92. Kwok, Y.K., Hwang, K., Song, S.: Selfish grids: Game-theoretic modeling and nas/psa benchmark evaluation. IEEE Tran. on Parallel and Distributing Systems 18(5), 1–16 (2007)
93. Lapin, L.: Probability and Statistics for Modern Engineering, 2nd edn. (1998)
94. Lee, Y.C., Zomaya, A.Y.: Minimizing energy consumption for precedence-constrained applications using dynamic voltage scaling. In: Proc. of the 9th IEEE/ACM International Symposium on Cluster Computing and the Grid CCGrid, Shanghai, China, pp. 92–99 (2009)
95. Lim, D., Ong, Y.S., Jin, Y.: Efficient hierarchical parallel genetic algorithms using grid computing. Future Generation Computer Systems 23(4), 658–670 (2007)
96. Liu, H., Abraham, A., Hassanien, A.: Scheduling jobs on computational grids using a fuzzy particle swarm optimization algorithm. Future Generation Computer Systems 26(8), 1336–1343 (2010)

97. Lorch, J., Smith, A.: Improving dynamic voltage scaling algorithms with pace. In: Proc. of the 2001 ACM SIGMETRICS International Conference on Measurement and Modeling of Computer Systems, pp. 50–61 (2001)
98. Lorenz, M., Wehmeyer, L., Dräger, T.: Energy aware compilation for dsps with simd instructions. In: Proc. of Languages, Compilers and Tools for Embedded Systems: Software and Compilers for Embedded Systems, LCTES/SCOPES 2002, pp. 94–101 (2002)
99. Lorpunmanee, S., Sap, M.N., Abdullah, A.H., Chompoo-inwai, C.: An ant colony optimization for dynamic job scheduling in grid environment. World Academy of Science, Engineering and Technology Bulletin 27 (2007)
100. Maheswaran, M., Ali, S., Siegel, H.J., Hensgen, D., Freund, R.F.: Dynamic mapping of a class of independent tasks onto heterogeneous computing systems. J. Parallel Distrib. Comput. 59, 107–131 (1999)
101. Mann, P.S.: Introductory Statistics, 7th edn. Wiley (2010)
102. Marcelloni, F., Vecchio, M.: Enabling energy-efficient and lossy-aware data compression in wireless sensor networks by multi-objective evolutionary optimization. Information Sciences 180, 1924–1941 (2010)
103. Meedeniya, I., Buhnova, B., Aleti, A., Grunske, L.: Architecture-Driven Reliability and Energy Optimization for Complex Embedded Systems. In: Heineman, G.T., Kofron, J., Plasil, F. (eds.) QoSA 2010. LNCS, vol. 6093, pp. 52–67. Springer, Heidelberg (2010)
104. Mejia-Alvarez, P., Levner, E., Mossé, D.: Adaptive scheduling server for power-aware real-time tasks. ACM Trans. Embed. Comput. Syst. 3(2), 284–306 (2004)
105. Mezmaz, M., Melab, N., Kessaci, Y., Lee, Y., Talbi, E.G., Zomaya, A., Tuyttens, D.: A parallel bi-objective hybrid metaheuristic for energy-aware scheduling for cloud computing systems. J. Parallel Distrib. Comput. (2011), doi:10.1016/j.jpdc.2011.04.007
106. Miao, L., Qi, Y., Hou, D., Dai, Y., Shi, Y.: A multi-objective hybrid genetic algorithm for energy saving task scheduling in cmp system. In: Proc. of IEEE Int. Conf. on Systems, Man and Cybernetics (ICSMC 2008), pp. 197–201 (2008)
107. Michalewicz, Z.: Genetic Algorithms + Data Structures = Evolution Programs. Springer (1992)
108. Milaneś-Montero, M., Romero-Cadaval, E., Barrero-González, F.: Hybrid multiconverter conditioner topology for high-power applications. IEEE Trans. on Industrial Electronics 58(6), 2283–2292 (2011)
109. Min, R., Furrer, T., Chandrakasan, A.: Dynamic voltage scaling techniques for distributed microsensor networks. In: Proc. IEEE Workshop on VLSI, pp. 43–46 (2000)
110. Nauck, D., Klawonn, F., Kruse, R.: Neuro-Fuzzy Systems. John Wiley & Sons (1997)
111. Norman, M.: Types of grid users and the customer-service provider relationship: a future picture of grid use. In: Cox, J. (ed.) Proc. of the UK e-Science All Hands Meeting, Nottingham, September 18-21, pp. 37–44 (2006)
112. Nowostawski, M., Poli, R.: Parallel genetic algorithm taxonomy, pp. 88–92 (1999)
113. Olivier, I., Smith, D., Holland, J.: A study of permutation crossover operators on the travelling salesman problem. In: Proc. of the ICGA 1987, Cambridge, MA, pp. 224–230 (1987)
114. O.L.: people.brunel.ac.uk/~mastjjb/jeb/info.html
115. Oracle, G.S.: www.oracle.com/technology/tech/grid/index.html
116. Page, J., Naughton, J.: Framework for task scheduling in heterogeneous distributed computing using genetic algorithms. AI Review 24, 415–429 (2005)
117. Pavlidis, N., Parsopoulos, K., Vrahatis, M.: Computing nash equilibria through computational intelligence methods. Journal of Computational and Applied Mathematics 175, 113–136 (2005)

118. Pedrycz, A.: Finite cut-based approximation of fuzzy sets and its evolutionary optimization. Fuzzy Sets and Systems 160(24) (2009)
119. Pedrycz, W.: Statistically grounded logic operators in fuzzy sets. European Journal of Operational Research 193(2), 520–529 (2009)
120. Pinel, F., Pecero, J., Bouvry, P., Khan, S.U.: Memory-aware green scheduling on multicore processors. In: Proc. of the 39th IEEE International Conference on Parallel Processing (ICPP), pp. 485–488 (2010)
121. Ranganathan, K., Foster, I.: Simulation studies of computation and data scheduling algorithms for data grids. J. of Grid Computing 1(1), 53–62 (2003)
122. Rayward-Smith, V.: Uet scheduling with unit interprocessor communication delays. Discrete Applied Mathematics 18(1), 55–71 (1987)
123. Reeves, C.: Landscapes, operators and heuristic search. Annals of Operations Research 86, 473–490 (1999)
124. Ritchie, G., Levine, J.: A fast effective local search for scheduling independent jobs in heterogeneous computing environments. TechRep, Centre for Intelligent Systems and Their Applications, University of Edinburgh (2003)
125. Rosettahome: `www.boinc.bakerlab.org/rosetta`
126. Roughgarden, T.: Stackelberg scheduling strategies. SIAM Journal on Computing 33(2), 332–350 (2004)
127. Rubio-Solar, M., Vega-Rodriguez, M., Perez, J., Gomez-Iglesias, A., Cardenas-Montes, M.: A fpga optimization tool based on a multi-island genetic algorithm distributed over grid environments. In: Proc. of the 8th IEEE International Symposium on Cluster Computing and the Grid (CCGRID 2008), pp. 65–72 (2008)
128. Rudolph, G.: Stochastic processes. In: Handbook of Evolutionary Computation B2.2 (1997)
129. Schaefer, R., Kołodziej, J.: Genetic search reinforced by the population hierarchy. In: FOGA VII, pp. 383–401. Morgan Kaufmann (2003)
130. SETI, h.: `www.setiathome.ssl.berkeley.edu`
131. Shelestov, A., Skakun, S., Kussul, O.: Intelligent model of user behavior in distributed systems. Int. J. on Information Theories and Applications 15, 70–75 (2003)
132. Shen, G., Zhang, Y.: A new evolutionary algorithm using shadow price guided operators. Applied Soft Computing 11(2), 1983–1992 (2011)
133. Shen, G., Zhang, Y.-Q.: A Shadow Price Guided Genetic Algorithm for Energy Aware Task Scheduling on Cloud Computers. In: Tan, Y., Shi, Y., Chai, Y., Wang, G. (eds.) ICSI 2011, Part I. LNCS, vol. 6728, pp. 522–529. Springer, Heidelberg (2011)
134. Solutions, H.G.: `h71028.www7.hp.com/enterprise/cache/125369-0-0-0-121.html`
135. Song, H., Liu, X., Jakobsen, O., Bhagwan, R., Zhang, X., Taura, K., Chien, A.: The microgrid: a scientific tool for modeling computational grids. Journal of Sci. Program 8(3), 127–141 (2000)
136. Song, S., Hwang, K., Kwok, Y.: Trusted grid computing with security binding and trust integration. J. of Grid Computing 3(1-2), 53–73 (2005)
137. Song, S., Hwang, K., Kwok, Y.: Risk-resilient heuristics and genetic algorithms for security-assured grid job scheduling. IEEE Tran. on Computers 55(6), 703–719 (2006)
138. Stadler, P.: Towards a theory of landscapes. In: Lopéz-Peña, R., et al. (eds.) Complex Systems and Binary Networks. Lecture Notes in Physics, pp. 77–163. Springer (1995)
139. Straffin, P.: Game Theory and Strategy. Mathematical Association of America Textbooks (1996)
140. Stutzle, T., Hoos, H.: Improvements on ant-system: Introducing max-min ant system. In: Proc. of the Artificial Neural Networks and Genetic Algorithms Conference, Wien, pp. 245–249. Springer (1996)

141. Subrata, R., Zomaya, A., Landfeldt, B.: Cooperative power-aware scheduling in grid computing environments. J. Parallel Distrib. Comput. 70, 84–91 (2010)
142. Talbi, E.G.: Metaheuristics: From Design to Implementation. John Wiley & Sons, USA (2009)
143. Tavakkoli-Moghaddam, R., Rahimi-Vahed, A., Mirzaei, A.: A hybrid multi-objective immune algorithm for a flow shop scheduling problem with biobjectives: weighted mean completion time and weighted mean tardiness. Information Sciences 177(2007), 5072–5090 (2007)
144. Tera-Grid, N.: `www.teragrid.org`
145. Thesen, A.: Design and evaluation of tabu search algorithms for multiprocessor scheduling. Journal of Heuristics 4(2), 141–160 (1998)
146. Veeramachaneni, K., Osadciw, L.: Swarm intelligence based optimization and control of decentralized serial sensor networks. In: Proc. of the IEEE Swarm Intelligence Symposium, pp. 1–8 (2008)
147. Venugopal, S., Buyya, R., Ramamohanarao, K.: A taxonomy of data grids for distributed data sharing, management, and processing. ACM Computing Surveys 38(1), 1–53 (2006)
148. Vose, M.: The Simple Genetic Algorithm. MIT press (1999)
149. Vose, M.D., Liepins, G.: Punctuated equilibria in genetic search. Complex Systems 5, 31–34 (1991)
150. Wang, L., Wei, L., Liao, X.-K., Wang, H.: *AT-RBAC*: An Authentication Trustworthiness-Based *RBAC* Model. In: Jin, H., Pan, Y., Xiao, N., Sun, J. (eds.) GCC 2004. LNCS, vol. 3252, pp. 343–350. Springer, Heidelberg (2004)
151. WebGrid, U.: `http://weboptserv.lsi.upc.edu/webgrid/`
152. Welch, V., Siebenlist, F., Foster, I., Bresnahan, J., Czajkowski, K., Gawor, J., Kesselman, C., Meder, S., Pearlman, L., Tuecke, S.: Security for grid services. In: Proc. of Int. Symp. High Performance Distributed Computing, HPDC-12 (2003)
153. Whitley, D., Rana, S., Heckendorn, R.: The island model genetic algorithm: On separability, population size and convergence. Journal of Computing and Information Technology 7, 33–47 (1998)
154. Wierzba, B., Semczuk, A., Kołodziej, J., Schaefer, R.: Hierarchical genetic strategy with real number encoding. In: Proc. of KAEiOG 2003, Łagów Lubuski, pp. 231–239 (2003)
155. Wu, C.C., Sun, R.Y.: An integrated security-aware job scheduling strategy for large-scale computational grids. Future Generation Computer Systems 26, 198–206 (2010)
156. Xhafa, F.: A hybrid evolutionary heuristic for job scheduling in computational grids. SCI, vol. 75, ch. 10. Springer, Heidelberg (2007)
157. Xhafa, F., Abraham, A.: Computational models and heuristic methods for grid scheduling problems. Future Generation Computer Systems 26, 608–621 (2010)
158. Xhafa, F., Alba, E., Dorronsoro, B., Duran, B.: Efficient batch job scheduling in grids using cellular memetic algorithms. Journal of Mathematical Modelling and Algorithms 7(2), 217–236 (2008)
159. Xhafa, F., Barolli, L., Durresi, A.: Batch mode schedulers for grid systems. International Journal of Web and Grid Services 3(1), 19–37 (2007)
160. Xhafa, F., Carretero, J.: Experimental study of ga-based schedulers in dynamic distributed computing environments. In: Alba, et al. (eds.) Optimization Techniques for Solving Complex Problems, ch. 24. Wiley (2009)
161. Xhafa, F., Carretero, J., Abraham, A.: Genetic algorithm based schedulers for grid computing systems. Int. J. of Innovative Computing, Information and Control 3(5), 1053–1071 (2007)

162. Xhafa, F., Carretero, J., Alba, E., Dorronsoro, B.: Tabu search algorithm for scheduling independent jobs in computational grids. In: Burguillo-Rial, J., Kołodziej, J., Nolle, L. (eds.) Computer And Informatics, Special Issue on Intelligent Computational Methods, vol. 28(2), pp. 237–249 (2009)
163. Xhafa, F., Carretero, J., Barolli, L., Durresi, A.: Requirements for an event-based simulation package for grid systems. Journal of Interconnection Networks 8(2), 163–178 (2007)
164. Xhafa, F., Gonzalez, J.A., Dahal, K.P., Abraham, A.: A GA(TS) Hybrid Algorithm for Scheduling in Computational Grids. In: Corchado, E., Wu, X., Oja, E., Herrero, Á., Baruque, B. (eds.) HAIS 2009. LNCS, vol. 5572, pp. 285–292. Springer, Heidelberg (2009)
165. Xie, T., Qin, X.: Enhancing Security of Real-Time Applications on Grids Through Dynamic Scheduling. In: Feitelson, D.G., Frachtenberg, E., Rudolph, L., Schwiegelshohn, U. (eds.) JSSPP 2005. LNCS, vol. 3834, pp. 219–237. Springer, Heidelberg (2005)
166. Yu, K.-M., Luo, Z.-J., Chou, C.-H., Chen, C.-K., Zhou, J.: A Fuzzy Neural Network Based Scheduling Algorithm for Job Assignment on Computational Grids. In: Enokido, T., Barolli, L., Takizawa, M. (eds.) NBiS 2007. LNCS, vol. 4658, pp. 533–542. Springer, Heidelberg (2007)
167. Zhang, G., Sun, J.: Grid intrusion detection based on soft computing by modeling real-user's normal behaviors. In: Proc. of the IEEE International Conference on Granular Computing 2006, pp. 558–561 (2006)
168. Zomaya, A.Y., Teh, Y.H.: Observations on using genetic algorithms for dynamic load-balancing. IEEE Tran. on Parallel and Distributed Systems 12(9), 899–911 (2001)
169. Zomaya, A.Y.: Energy-aware scheduling and resource allocation for large-scale distributed systems. In: 11th IEEE International Conference on High Performance Computing and Communications (HPCC), Seoul, Korea (2009)

Index

Printed by Publishers' Graphics LLC
BT20121030.19.18.126